CONSTRUCTION ESTIMATING AND COSTING

Frank W. Helyar, F.R.I.C.S.

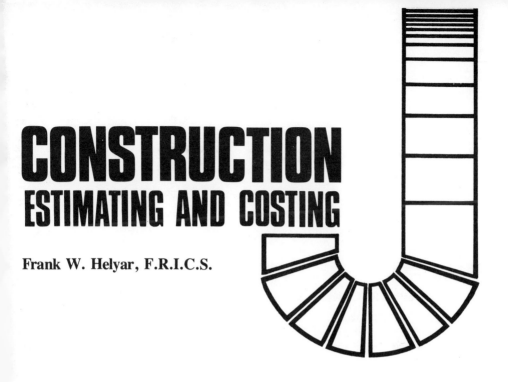

McGRAW-HILL RYERSON LIMITED

Toronto Montreal New York St. Louis San Francisco
Auckland Beirut Bogotá Düsseldorf Johannesburg
Lisbon London Lucerne Madrid Mexico New Delhi
Panama Paris San Juan São Paulo Singapore Sydney
Tokyo

CONSTRUCTION ESTIMATING AND COSTING

Copyright© McGraw-Hill Ryerson Limited, 1978. All rights reserved. No part of this publication may be reproduced, stored in a retrieval system, or transmitted, in any form, or by any means, electronic, mechanical, photocopying, recording, or otherwise, without prior written permission of McGraw-Hill Ryerson Limited

Hardcover ISBN: 0-07-082610-2

1 2 3 4 5 6 7 8 9 10 HR 7 6 5 4 3 2 1 0 9 8

Printed and bound in Canada

Canadian Cataloguing in Publication Data

Helyar, Frank W.
 Construction estimating and costing

Bibliography: p.
Includes index.
ISBN 0-07-082610-2

1. Building - Estimates. I. Title.

TH435.H44   692'.5   C78-001506-1

Table of Contents

Preface   v

Acknowledgements   xi

PART 1   CONSTRUCTION COST PLANNING AND COST CONTROL

Chapter 1   THE NEED FOR COST CONTROL   1
Chapter 2   FACTORS INFLUENCING COST   7
Chapter 3   ESTIMATING METHODS   23
Chapter 4   ELEMENTAL COST ANALYSIS   33
Chapter 5   COST PLANNING   49
Chapter 6   COST CONTROL   57

PART 2   ECONOMICS OF DEVELOPMENT

Chapter 7   FEASIBILITY STUDIES   65
Chapter 8   FINANCING   75
Chapter 9   TIME AND MONEY   80
Chapter 10   YIELD ANALYSIS   94
Chapter 11   TAXATION   117
Chapter 12   LIFE-CYCLE COSTING   125

APPENDIX A   157
APPENDIX B   161
APPENDIX C   173
APPENDIX D   227
BIBLIOGRAPHY   228
INDEX   229

Preface

The genesis of this book was a series of notes prepared for some lectures given in the Construction Option course at the University of Waterloo. It soon became apparent that they had expanded themselves to such an extent that they might form the basis for a book and this is the outcome.

The lectures were on the subject of Construction Economics, and were mostly concerned with what is now Part One of this book. They were given to a small select group of people who were studying to become project managers but I hope the book will be of interest to architects, engineers, quantity surveyors, contractors, developers and in fact to anyone who is connected with the construction industry.

As a quantity surveyor I hope this book will indicate to other members of the construction industry, and to potential quantity surveyors, some of the scope of training and experience required by quantity surveyors which makes them more than just construction estimators.

Quantity surveying is not yet the recognized profession in North America that it is in the United Kingdom, parts of Europe and in many of the Commonwealth countries. However, even in the United Kingdom it is a comparatively recent profession and although there are references in both the Bible[1] and Shakespeare[2] to work which might be construed as normally being performed by quantity surveyors it was really only in the eighteenth century that the quantity surveyor started to become identified as a separate entity, and only in the nineteenth century that he became a recognized professional.

Prior to the seventeenth century in England the design of buildings was generally confined to the drawing of a ground plan or "platte", usually by a surveyor but sometimes by one of the craftsmen. The importance of the embryonic architect of those days can be judged by the following adjoining entries which can be found in the *Black Books of Lincoln's Inn* for 1561:

[1] Luke 14:28-30
[2] Henry IV Part II

1. Paid Potter the Bricklayer for drawing the plans for building the Inne 4s 0d
2. Received a fine from Fleming for having his beard too long 3s 4d

If the building were of some importance the surveyor, for an additional fee, would also provide an elevation or "upright."

A copy of the plan was given to the craftsmen, each of whom worked in his own trade with his own men, and they would calculate their labour costs and send the bill to the building owner rather like a present day cost plus project. On smaller jobs they usually sent their bills after the work was completed, but on larger jobs they were frequently paid a cash advance to be adjusted on completion. The owner, in addition to paying for the labour, would supply the workmen with the necessary materials to work with.

If the building were important and large enough the work would be under the general coordinating supervision of a building surveyor who was usually a master mason or carpenter. He was a man of some standing, and if he were sufficiently well known he might have several buildings under his control. He would visit each in turn, laying out the work and doing some of the more intricate details himself, while his apprentices and the workmen who would complete the job looked on or worked under his guidance.

Inigo Jones is credited with importing from Italy the concept of having an architect design and supervise the building from beginning to end, and with it came the Italian Method of having a single contractor to superintend the work, and pay the craftsmen. His Banqueting Hall in London's Whitehall, built in 1620, was probably the last of the great buildings to be built by individual craftsmen, although the old practice continued for lesser buildings into the nineteenth century, while his Lincoln's Inn Chapel, designed in the same year, was built by a single contractor. At this time too the designation architect began to replace that of surveyor, although to this day some of the London guilds still retain a surveyor to the guild, a position usually occupied by an architect.

Despite the changes brought about by Inigo Jones it was still customary for the contractor to be paid in advance for at least some of the work, and for the building owner to supply the materials. Payments were made during the course of construction on a weekly or monthly basis, either on the cost of wages expended or by measurement of the amount of work done.

Where payment was to be made on measurement rather than on the actual wages paid, unit rates were agreed upon between the workmen or the contractor and the owner, usually with the architect acting as the owner's representative. The architect was responsible for agreeing measurements and would normally employ a measurer or measuring surveyor to do so. However, some architects used clerks or pupils to do the work. No doubt the clerks and pupils were not particularly interested in this type of work because in time the workmen became dissatisfied with this arrangement and employed their own measuring surveyors to look after their interests.

PREFACE

The job of a measuring surveyor was not at that time considered to be a very elevated one, and many of the surveyors had a poor reputation for accuracy and honesty. There was no formal training for them, and very little in the way of books to guide them. In 1663 Sir Balthazar Gerbier produced his book *Counsel and Advice to all Builders* which contained long schedules of prices but was a superficial and padded work. Stephen Primatt published his *City and Country Purchaser and Builder* in 1667, the year following the Great Fire of London. He gave as his reason for doing so the fact that:

> Many of those wanting to build are compelled to trust to the Fidelity of Workmen and Surveyors who have been observed to make a Harvest in the Citie Ruines and combine together to take excessive Rates for their work which hath discouraged many . . . The purchaser is oft-times over-reached by the seller and thereby undone.

Prior to this the Carpenters' Rule for the easy measuring of "superficies and solids" had been devised, and Leonard Digges had published *A Book Named Tectonicon* in 1556 "for the exact measuring, and speedie reckoning all manner of Land, Square, Timber, Stone, Steeples, Pillars, Globes etc." with an instrument known, ominously, as a "profitable staff."

Another reason for the poor reputation enjoyed by measuring surveyors could have been the transient nature of the business. Keat's *London Directories* lists the unusual name of Ezekiel Delight as Oil and Colourman in 1779, as a builder shortly afterwards, and as a surveyor in 1794.

Two events occurred at the beginning of the nineteenth century which radically changed the construction industry. The first was that, due to the Napoleonic wars, the government decided "on the score of economy and for the sake of despatch" all building projects should be under the control of one contractor. This was, of course, already being done on major projects but it now became mandatory for projects of all sizes.

The second event was a demand for open competitive tendering. Whereas previously the total cost of the job was only known when the work was finished, owners now required that the contractors give a price before work started. One of the effects of this was that measurements had to be taken from the drawings before construction started rather than on the site after the work was built, and as a result many of the old measuring surveyors, at least the more inefficient ones, soon found themselves without work.

A book written by John Noble in 1836 describes the new system.

> An excellent system has been recently adopted by the Craft, in selecting a few respectable builders to tender for the performance of the work; then appointing one surveyor and the employer another, to ascertain the quantities together and to bring the same into blank account; so that each party proceeds to value the items upon the same accurate data, on which he founds his tender, and this mode is beneficial to them and the public.

At first this system required two or more surveyors to measure the quantities, a carry-over of the old practice where both parties measured the work for progress payments. However, as confidence in the skills of the surveyor grew it came to be realized that this was rather inefficient and time-wasting, and the contractors agreed to accept the quantities measured by one surveyor who was appointed and paid by all the contractors who were submitting a tender, and the requirement of a surveyor measuring on behalf of the owner was dropped. Later, to preserve secrecy in tendering, it was agreed that the architect on behalf of his client should appoint the surveyor. A relic of the old system survived until shortly after World War II when the successful contractor was required to pay the quantity surveyor's fee, having included the amount in his tender.

The schedules of labour and materials prepared by the surveyor for the contractors to tender on became known as Bills of Quantities, and the surveyor changed from being a measuring surveyor to being a quantity surveyor. Together with the change in name, quantity surveying became an independent profession about the middle of the nineteenth century, although even as late as 1880 quantities and competitive tendering were described as "the modern system of contracting and tendering for work."

The profession has evolved gradually over the past 150 years, with many major changes taking place in the last 50 years. When an independent quantity surveyor is measuring quantities for contractors to price it is essential that the contractors know exactly how they were measured, what is included and what is not included in the descriptions of the work items, and so on. Quantity surveyors therefore established guidelines for the use of the contractors which might vary from firm to firm and from district to district. In 1922 this situation was regularized by the publication of uniform guidelines applicable throughout England and Wales, known as *The Standard Method of Measurement of Building Works* and published jointly by the Royal Institution of Chartered Surveyors and the National Federation of Building Trades Employers. This document is now in its sixth edition, major revisions having taken place with each new edition, and since the 1950s has also been accepted in Scotland. Other publications covering engineering works and small building works are now also available.

A bill of quantities is broken down into trades to assist the contractor when he is preparing his tender. The general contractor will price all those items with which he is concerned and send a copy of the quantities to all the trades from whom he wants to receive a price. None of them will measure any of the quantities but will accept them as being accurate. Competition is therefore on price rather than on errors in quantities. At tender closing only lump sum prices are submitted, but the lowest bidder will be required to submit a completely priced bill of quantities to the quantity surveyor within 24 or 48 hours for checking. Any arithmetical or other errors are adjusted and, provided his tender is still the lowest, he is likely to be awarded the contract. If his corrected price puts him above the next bidder he is given the option of standing by his tender or making the correction. Unless the error is a major one he will usually stand by his tender since he knows that making the correction will lose him the job. When the tender price has been settled and the contract signed the priced bill of quantities becomes one of the contract documents, to be used for adjusting the cost of changes in the work during construction. It should be noted that only the low tenderer is required to reveal his unit prices, and then only to the quantity surveyor. The priced bill of quantities should not be made available to the architect or the building owner.

As has already been mentioned, the unit prices in the bill of quantities are used for adjusting the cost of changes in the work during construction. They are also used for correcting any errors in the quantities which may subsequently be found. The contractor accepts the quantities as being accurate at the time of tendering, but if he finds out later that an item was under-measured he is entitled to claim an extra to correct the error. He is also required to submit a credit if he finds that an item was over-measured. Since the error is adjusted at the unit price contained in the bill of quantities the owner should have no hesitation in accepting it because in theory, at least, the low tender would have been higher by that amount if the error had not been made.

Quantity surveyors have progressed a long way from the old measuring surveyors of the eighteenth century. Besides preparing bills of quantities they are also responsible for negotiating the cost of changes in the work and settling the final account, and for valuing the progress payments although it is the architect's responsibility to issue the certificate on the recommendation of the quantity surveyor. Because of the cost expertise they have acquired, quantity surveyors have become not only the technical accountants of the construction industry, but are also the consultants used to advise the building owner on probable costs of construction in the pre-tender stages, using cost planning and cost control techniques developed since the Second World War. Many firms now also rely very heavily on computers for all aspects of their work.

NORTH AMERICA

In North America the quantity surveying profession has not evolved in the same way as it has in Britain. While Australia, New Zealand and other parts of the Commonwealth have adopted the British system with some local modifications, and even parts of Europe have changed to the bills of quantities system in recent years, North America remains virtually the last English-speaking area where quantity surveying is not a generally recognized profession.

Because the contractors and sub-contractors have to measure their own quantities when tendering a project, quantity surveyors are generally found on their staffs rather than as independent consultants. The last 20 years has, however, seen the emergence of a few firms of independent quantity surveyors who offer pre-tender cost advice to architects and building owners, using the cost planning and cost control techniques developed in Britain. They also offer other services usually associated with quantity surveying such as expert evidence, replacement costs, mortgage appraisals and financial supervision on behalf of investors.

In 1959 the Canadian Institute of Quantity Surveyors was formed in Toronto. Its principal aims are to provide a means of communication between quantity surveyors, to act as a spokesman for its members, and to act as an examining body for those who wish to become quantity surveyors. This last role has required the establishment of a course of studies and the investigation of various educational institutions together with the offer of advice and encouragement to potential members.

In recent years, because Canada is such a large country, it was decided to create autonomous provincial associations which are affiliated with the Canadian Institute of Quantity Surveyors. Currently they are the Quantity Surveyors Society of British Columbia, the Ontario Institute of Quantity Surveyors and the Quantity Surveyors of Quebec Incorporated. Outside these provinces the Canadian Institute of Quantity Surveyors is still the body primarily governing the affairs of quantity surveyors.

The Ontario Association of Consulting Quantity Surveyors was founded in 1974 as an association of consulting firms which supports the aims and objectives of the Canadian Institute of Quantity Surveyors but which is intended to look after the interests of those firms in private practice in Ontario.

Canadian quantity surveyors have found a place for themselves in the construction industry over the past twenty years. They may not prepare bills of quantities as their counterparts in other parts of the world do, but they are able to provide all the other services offered by quantity surveyors elsewhere.

Acknowledgements

I wish to acknowledge with thanks the help provided to me in the preparation of this book by the following people:
My wife Joyce and Doris G. MacLean for typing the manuscript.

Philippa Dunn of McGraw-Hill Ryerson who edited the final draft and made sense out of what had been obscure.

G.I.M. Young and A. G. Williams of Stewart, Young & Mason for their constructive criticism of the first draft.

Hugh D. Thomas of the Waterfront Development Corporation, Limited, Halifax, Nova Scotia, for help with the list of capital costs in chapter seven.

R. J. Evans of Edgecombe Realty Limited for much useful information on property management and the operating expenses of buildings.

D. C. Crawford of A.E. LePage Investment and Professional Services Company who first introduced me to the natural mortgage constant mentioned in chapter eight.

J. Strung of Strung Real Estate Limited from whom I learned much about yield analysis.

W. G. Ralph of J. Clare Wilcox & Co., Chartered Accountants, who provided all the information on which chapter eleven is based.

I take the view, and always have done, that if you cannot say what you have to say in twenty minutes, you should go away and write a book about it.

Lord Brabazon of Tara

PART 1

Construction Cost Planning and Cost Control

CHAPTER 1

The Need for Cost Control

In the past, when an architect was asked for an estimate on a building he was engaged to design he usually worked it out by measuring the floor area or the cube of the building. He would then multiply the result by a unit price obtained from previous projects to give a total estimated cost. Since all buildings are different he had to use his judgement in the selection of the unit price, coupled perhaps with a certain amount of intuition; and he would often confirm the result with the help of a friendly contractor.

This may have been satisfactory a hundred years ago when construction techniques were comparatively simple and not subject to major changes from year to year and from project to project, and when cost inflation was minimal. Today however building requirements are becoming more complicated, and with all the new materials and construction techniques which are now readily available, together with the extreme inflation which has taken place over the past few years, there is a greater opportunity for costs to get out of control in the pre-tender period. It means that nineteenth century estimating techniques are no longer valid for the twentieth century. Not only because they were intended simply as a means for estimating construction costs and no longer give very accurate results, but also because they cannot be used to help control costs.

The need for cost control will be viewed in different ways by the various parties to a construction project: the building owner, the design consultants and the contractor.

THE BUILDING OWNER

It almost goes without saying that the building owner is greatly concerned with the cost of his proposed building. Besides wanting to be assured that he is getting the best value for his money, the prospective owner wants to be certain that the money set aside for the building at the conceptual stage is sufficient when tenders are opened. For a prospective house owner the decision to build is probably the most important financial decision of his life, requiring an initial estimate which is as accurate as possible. Similarly, if the proposed building is a speculative office building for which the developer has allocated an estimated construction cost in the feasibility study, an increase of only three or four per cent in the actual construction cost could mean the difference between success and disaster. Although the situation might not always be so critical most prospective owners, whether private clients or government departments, have a very real concern that costs stay within the budget.

Despite the fact that cost control is so crucial to the building owner it is not unknown, on a stipulated sum tender call, for tenders to be opened and found to be higher than the owner or his consultants had expected. This situation leaves the owner with four options:

1) He abandons the project. This means that he will have been put to some expense without having a building to show for it, while his consultants will have suffered a loss in reputation as well as a portion of their fee.

2) He has the building redesigned to meet the budget. This is a quite frequent but very unsatisfactory solution since a design which has been carefully thought out over a period of several months can only suffer if it has to be radically changed within a few days. The owner is upset because of the disruption to his schedule and possible extra expense caused by the delay and the design team is put to additional work and expense without any increase in its fee.

As a matter of interest in the October 1976 *Standard Form of Agreement between Client and Architect* issued by the Ontario Association of Architects the architect is required to prepare a final estimate of cost just prior to calling tenders. This estimate is to be accepted by the client and if the low tender exceeds the final estimate by an agreed percentage the architect must redesign the project to bring the cost down to an acceptable level. The architect's fee is then based on the final cost of the work, unless the changes he makes result in a cost which is less than his final estimate, in which case the architect's fee is based on his final estimate.

3) He fires his consultants and brings in new ones, who, he hopes, will do a better job of cost control. Again the loss of fees and, in particular, the original consultants' loss of reputation can cause irreparable harm.

4) He finds money to cover the extra expense. In this case he may have an architectural monument (or a monument to an architect) when the building is complete, although he may feel that despite his irritation over the additional expense the resulting building was well worth it. After all, the financial

problems of the building owner are rarely considered when admiring the architectural qualities of a fine building. In this connection it might be noted that the cost of Buckingham Palace exceeded its architect's estimate, and the same may have been true of the Parthenon, and is certainly true of many other notable buildings.

Regardless of which course of action he has to take, the owner, because his consultants have taken insufficient care about the costs, will be put to additional expense; either directly or indirectly because of delay in construction, and time has a measurable value to most owners.

Where the building is to be built under a management contract a construction manager is appointed and the detailed design proceeds in sequence with the calling of trade package tenders as the building is being built. In this case (known as fast-track in the United States) the owner's options are reduced from four to two when costs start to exceed the budget. He can hardly abandon the project if it is already half built, and it is unlikely that he would fire his consultants part way through the project. All that he can do is to have changes made in the design to reduce the cost of any trade package which is too high, or find the extra money. Usually, however, with this approach unless the budget is outrageously low there is sufficient flexibility to allow for those trades which have been over-estimated to compensate for those which have been under-estimated.

Although cost control is normally of great concern to the building owner, there are those rare occasions when cost is only of secondary concern. Such situations arise when there is a difference between cost and value. Cost and value are usually the same but in some instances there may be a difference between the two.

Many books and articles have been written on the theory of value: whether the value of an object depends on a personal viewpoint, or on the price someone is prepared to pay for it, or on the power it bestows on its owner. Value will be discussed in more detail in later chapters since it is the criterion by which the economic feasibility of a project is judged. However, it might be useful at this point to look at some examples where cost and value might appear to differ depending on the viewpoint of the building owner.

If someone were to build a house at a cost of $250,000 he would probably feel this sum represents its value otherwise he wouldn't have spent that much on it. If, however, he tried to sell the house on the open market and could only get $150,000 for it, the value to anyone else is obviously far less than the cost.

Similarly, a high-quality building in a poor locality, or a building which is out of scale with its community, is likely to have a value which is less than its cost. In the former case it sometimes happens that the construction of a superior building in a run-down community will spark general redevelopment in the area and as a result the value will increase, although not necessarily enough to meet its cost. In the latter case a 55 - storey building in a small town would be a very unwise investment since its cost would far exceed its value.

While these examples give instances of cost apparently exceeding value, the reverse can also be true. The erection of a building on a raw piece of land is always intended to increase the value of the property, but an inexpensive parking structure on an existing parking lot for example could increase the income of the property, and hence its value, to a point well beyond the cost.

Where the building owner's principal business is something other than the ownership of the building, and the building is therefore of only secondary importance to him he may be prepared to increase the construction cost beyond what might normally be considered reasonable in order to increase the value of his business. If for example, a major department store were to burn down, or if a manufacturer wanted to get a new product on the market as soon as possible, they would probably be willing to pay a premium for the construction because the value is in having sales or manufacturing space available at the earliest possible moment, not in the physical value of the building.

The relationship between cost and value is really in the realm of the appraiser, and sometimes it can become quite complicated. As an example, a large office in London, England, was built in 1963 and remained completely empty until 1974. It has been suggested that the reason it was not rented was because the developer calculated that in a period of increasing rental rates, and with no taxes on unoccupied buildings, its value was increasing over the years at a greater rate than it would if it were rented. This is because if space were rented at, say, $90.00 per m^2 for a five-year period, any valuation based on income during that five-year term would use the $90.00 figure in the calculations. If, in the year following the lease agreement, the same space could have commanded a rent of $110.00 per m^2 because of inflation this would have to be ignored in the valuation, but with the space unrented the $110.00 figure could be used. Thus, theoretically at least, the value increased at a higher rate because the building was empty.

THE DESIGN CONSULTANTS

All that has been said about the building owner's concern with costs applies equally to the design consultants because they should be aware of the owner's concern and be prepared to respond accordingly. They should be able to assess the effect of any design changes on the cost so the client can be kept informed of his current financial commitments.

As has already been indicated the design consultants' reputation can suffer if they pay no heed to cost control. Moreover, any professional office needs at least to cover its costs if it is to remain in business, and to do so it should operate its own internal cost control system, allocating portions of the anticipated fee to design, working drawings, construction supervision and so on. Since professional fees are usually based on a percentage of the construction cost, any error in forecasting that cost can seriously upset the fee budget and thereby affect the firm's profitability.

THE CONTRACTOR

The contractor, although he is not usually consulted during the design of a building, has a rather indirect, but nevertheless very real, interest in pre-tender cost control. His interest in cost control during construction, on the other hand, is very direct and sometimes at odds with the owner's interests.

The contractor and his sub-contractors are in business to make money. Tendering is a major overhead expense and the contractor's estimating department can represent 40% to 50% of his head office overhead. He therefore likes to know that if he submits the lowest tender he will be building the project, while if his price is too high he can pass on to other tenders. If, because of insufficient cost control by the owner and his consultants, he has to re-tender on a modified design, a disproportionate amount of his overhead has to be expended on the project. Worse still, if he is low bidder and the project is cancelled because it is too expensive, the contractor knows he has been wasting his time and will be wary about submitting tenders to that owner or his consultants in the future.

Further, when a building has been redesigned to meet the budget there is a good chance that small items have been overlooked, resulting in a number of changes in the work during construction. The contractor likes his building operation to proceed smoothly with a minimum of changes because a smooth operation is more profitable, and this is an additional reason for the contractor to have an interest in proper pre-tender cost control. Contrary to a widely held belief, contractors do not make money on changes unless they are very substantial.

While tenders which are over the budget can cause trouble for the contractor, tenders which are well under the budget can be a source of irritation. If proper cost control had been exercised the contractor would have been bidding on a building which more closely resembled the budget with a consequently larger profit, and since his business is to make money out of construction it is bound to irk him when he sees potential profits disappearing.

When a contractor is putting together a tender for a stipulated sum contract he estimates the cost of his own work, and then adds to it the lowest bid for each of the sub-trades, the amounts of any cash allowances the architect has included in the specifications, and his overhead and profit. The co-ordination required is formidable, particularly since most of the work is now done by sub-contractors and the co-ordination has to be done in the last few minutes before the tender closes. It might help reduce the confusion a little if the design consultants were to bear the following in mind when they are calling for tenders:

1. Whenever possible a selected list of tenderers should be used. A construction project always runs more smoothly when it is being handled by a contractor who has submitted a reasonable price and is trusted by the owner and his consultants. An open list can result in a contractor whom nobody has ever heard of, who has submitted a low price because he left out part of the work in his estimate and who, in attempts to make up his loss, will cut corners and make life difficult for everyone.

2. Complicated tender forms should be avoided. If numerous unit prices and alternates are essential to a proper analysis of the tender, they should be asked to be submitted 24 or 48 hours after tendering. The contractors can then give them the consideration they deserve.

3. If addenda are required the last one should go to the contractors at least three days before the tender closes. An addendum may appear simple, but for the contractor it may require a considerable revision to his estimate. It should not be too much to ask that when the contract documents have been carefully assembled by the consultants over a period of some weeks or months the contractor should be given time to prepare a proper estimate without being inundated with addenda at the last minute.

4. Closing a tender the day after a holiday or a weekend means that the contractors' estimators won't have a holiday or a weekend. The best time to close a tender is around the middle of the week, late in the afternoon, and preferably when no other tenders are being closed.

All parties to a construction project are affected when pre-tender cost control is not applied, either directly as in the case of the building owner, or indirectly as in the case of the consultants, the contractor and the sub-contractors, all of whom rely on the building owner for their income.

CHAPTER 2
Factors Influencing Cost

THE ECONOMY

Construction costs are influenced by factors at three levels in the economy: at the national or international level, at the industry level, and at the level of the individual building. The first two levels are really in the realm of the economist but anyone involved with construction costs should have some knowledge of their effect.

Figures published by Statistics Canada and shown in Tables 1 and 2 indicate that in 1976 construction, including new construction and repairs, represented approximately 17% of the gross national product but, perhaps more importantly, it represented about 73% of the gross fixed capital formation. This means that of the total amount expended by government and industry on construction, machinery and equipment, 73% went to construction. The construction industry is therefore an important segment of the national economy and any changes in the economy can substantially affect the industry.

When money becomes scarce and the future is not being viewed with any great optimism by entrepreneurs there is a tendency for the private sector to reduce capital formation. Since construction forms such a large part of capital formation the volume of construction will drop, tending to lead to a reduction in tender prices as contractors start looking desperately for work and cut their profit margins to obtain it. This is inevitably followed by an increase in the number of bankruptcies and a general depression in the construction industry.

On the other hand, a period of inflation will increase the cost of wages and materials, construction costs will rise, and this may also lead to a reduction in the volume of construction because it has become too expensive, particularly if interest rates and land costs have also become excessive.

The Economic Council of Canada's report *Toward More Stable Growth in Construction* gives an admirable review of instability in the construction

CONSTRUCTION ESTIMATING AND COSTING

industry and how it might be ameliorated. On the subject of the construction industry in the economy it states:

> Construction instability is an integral part of the larger process of economic expansion and contraction occurring locally, regionally and nationally. Since construction firms primarily are marketing highly specialized site-preparation and assembly skills, they carry no inven-

Table 1: The Gross National Product, Fixed Capital Formation and Construction Activity in Canada in Current Dollars

Year	Gross National Product	Fixed Capital Formation	Total Construction Activity	Total Construction Activity as percentage of	
				GNP	FCF
	$000 000			%	
1964	50 280	11 205	8 662	17.23	77.30
1965	55 364	13 179	9 929	17.93	75.34
1966	61 828	15 361	11 235	18.17	73.14
1967	66 409	15 628	11 621	17.50	74.36
1968	72 586	15 754	12 213	16.83	77.52
1969	79 815	17 232	13 207	16.55	76.64
1970	85 685	18 015	13 781	16.08	76.50
1971	94 115	20 474	15 865	16.86	77.49
1972	104 669	22 598	17 289	16.52	76.51
1973	122 582	27 203	20.174	16.46	74.16
1974	144 616	33 597	24 693	17.07	73.50
1975	161 132	39 230	28 133	17.46	71.71
1976	184 494	43 182	31 449	17.05	72.83

Source: Statistics Canada

Table 2: The Gross National Product, Fixed Capital Formation and Construction Activity in Canada in 1971 Dollars

Year	Gross National Product	Fixed Capital Formation	Total Construction Activity	Total Construction Activity as a percentage of	
				GNP	FCF
	$000 000			%	
1964	65 610	14 549	12 016	18.31	82.59
1965	69 981	16 259	13 047	18.64	80.24
1966	74 844	18 015	13 900	18.57	77.16
1967	77 344	17 942	13 806	17.85	76.95
1968	81 864	17 964	14 321	17.49	79.72
1969	86 225	18 850	14 656	17.00	77.75
1970	88 390	18 904	14 605	16.52	77.26
1971	94 115	20 474	15 865	16.86	77.49
1972	99 680	21 612	16 367	16.42	75.73
1973	106 845	23 997	17 383	16.27	72.44
1974	110 293	25 231	18 018	16.34	71.41
1975	110 975	25 848	18 193	16.39	70.38

Source: Statistics Canada

tories. Instead, they respond – as do service agencies – to the immediate demands of other sectors of the economy: to business demands for additional industrial or commercial plant capacity; to government spending on schools, hospitals, energy facilities, roads, sewers and other special structures; and to family requirements for new or replacement housing.

At the same time it points out that while Canada's investment was more stable than the United States in the period 1948-70, "construction in Canada contributed much more to the instability of the economy."

THE CONSTRUCTION INDUSTRY

While the national economy has an effect on construction costs, the structure of the industry itself also has an effect. The availability and cost of labour and, to a lesser extent, the availability and cost of materials are both within the orbit of the industry, and consequently will affect the overall cost of construction. But the industry also has its own peculiarities which make it quite unlike any other type of manufacturing industry.

The creation of capital goods by the construction industry is partly a production process, converting raw and semi-finished materials into their final form, and partly an assembly process, assembling manufactured components into a completed product. To do this production is undertaken on an open site rather than in an enclosed factory and this poses unique problems of organization and management. It also makes production dependent to a certain extent on the weather, and it requires a high degree of mobility of labour and materials. As the Economic Council of Canada puts it:

> Teams of construction contractors, tradesmen and other specialists are continually organizing in patterns appropriate to the technical and commercial requirements of each project. When the job is done, they depart to join teams on other projects. This feature of temporary arrangements is characteristic of the construction industry and, in addition to cyclical and seasonal swings in demand, is an important source of insecurity for the participants.

The construction industry is largely sheltered from international competition in the goods it produces. Although the materials it uses may be imported, and some construction materials are exported, the finished products remain in a fixed location unlike the products of most other manufacturing industries.

Construction is a widely dispersed and compartmentalized industry, with a wide range of products being produced for its clients by a large number of firms across the country. The products themselves are individual in character, requiring their own plans and specifications and containing an immense

Table 3: Characteristics of the Non-Residential Construction Industry in Canada, 1974

Description	Size of Work Undertaken					
	$0— 249 999	$250 000— 499 999	$500 000— 999 999	$1 000 000— 1 999 999	$2 000 000— 9 999 999	$10 000 000 and over	Total
Type of Company:							
Sole owner	135	16	8	—	2	—	161
Partnership	29	7	1	6	—	—	43
Incorporation	505	284	269	201	154	12	1 425
Total	669	307	278	207	156	12	1 629
Type of Construction:							
Industrial	170	95	97	84	50	4	500
Commercial	316	133	97	74	63	7	690
Institutional	106	62	66	40	41	1	316
Other	77	17	18	9	2	—	123
Type of Work:							
New	540	264	256	201	151	12	1 424
Repairs	129	43	22	6	5	—	205
Value of Construction*							
Industrial	34 127	59 588	123 312	198 247	356 397	136 829	908 500
Commercial	69 370	93 965	138 242	242 824	623 969	285 208	1 453 578
Institutional	36 284	50 115	115 904	129 562	348 289	84 884	765 038
Other	20 020	23 658	43 492	58 092	127 662	13 661	286 585
Total	159 801	227 326	420 950	628 725	1 456 317	520 582	3 413 701
Value of Work*							
New	129 509	187 487	373 424	569 252	1 374 385	518 632	3 152 689
Repairs	30 292	39 839	47 526	59 473	81 932	1 950	261 012

*Thousands of Dollars

Adapted from information published by Statistics Canada in *Catalogue 64-207 Annual*. The non-residential general building contracting industry.

Table 4: Characteristics of the Residential Construction Industry in Canada, 1974

Description	Size of Work Undertaken							Total
	$0– 99 999	$100 000– 249 999	$250 000– 499 999	$500 000– 999 999	$1 000 000– 1 999 999	$2 000 000– 9 999 999	$10 000 000 and over	
Type of Company:								
Sole owner	615	245	51	8	2	—	—	921
Partnership	52	51	15	3	5	—	—	131
Incorporation	490	683	463	259	144	101	19	2 159
Total	1 157	979	529	270	151	106	19	3 211
Type of Construction								
Single Family	1 135	941	503	253	137	90	16	3 075
Apartments	22	38	26	17	14	16	3	136
Type of Work:								
New	1 039	945	514	265	148	106	19	3 036
Repairs	118	34	15	5	3	—	—	175
Value of Construction:*								
Single Family	94 190	229 911	263 133	257 160	284 593	556 448	550 989	2 236 424
Apartments	5 772	14 409	17 475	37 426	43 170	98 592	48 171	265 015
Other	3 219	11 272	12 864	15 992	16 706	44 591	33 855	138 499
Total	103 181	255 592	293 472	310 578	344 469	699 631	663 015	2 639 938
Value of Work:*								
New	92 482	242 157	280 000	301 502	336 702	698 115	663 015	2 583 973
Repairs	10 699	13 435	13 472	9 076	7 767	1 156	—	55 965

*Thousands of Dollars

Adapted from information published by Statistics Canada in *Catalogue 64-208 Annual. The residential building contracting industry 1974.*

Table 5: Characteristics of the Residential and Non-Residential Construction Industry in Canada, 1974

Description	$0–249 999	$250 000–499 999	$500 000–999 999	$1 000 000–1 999 999	$2 000 000–9 999 999	$10 000 000 and over	Total
Type of Company:							
Sole owner	995	67	16	2	2	—	1 082
Partnership	132	22	4	11	5	—	174
Incorporation	1 678	747	528	345	255	31	3 584
Total	2 805	836	548	358	262	31	4 840
Type of Construction:							
Industrial	170	95	97	84	50	4	500
Commercial	316	133	97	74	63	7	690
Institutional	106	62	66	40	41	1	316
Residential	2 136	529	270	151	106	19	3 211
Other	77	17	18	9	2	—	123
Type of Work:							
New	2 524	778	521	349	257	31	4 460
Repairs	281	58	27	9	5	—	380
Value of Construction:*							
Industrial	34 127	59 588	123 312	198 247	356 397	136 829	908 500
Commercial	69 370	93 965	138 242	242 824	623 969	285 208	1 453 578
Institutional	36 284	50 115	115 904	129 562	348 289	84 884	765 038
Residential	353 115	293 529	312 012	352 817	689 368	608 461	2 609 302
Other	25 678	23 601	42 057	49 745	137 926	38 214	317 221
Total	518 574	520 798	731 527	973 195	2 155 949	1 153 596	6 053 639
Value of Work:*							
New	464 149	467 486	674 925	905 955	2 072 501	1 151 647	5 736 663
Repairs	54 425	53 312	56 602	67 240	83 448	1 949	316 976

* Thousands of Dollars
Adapted from Tables 3 & 4

variety of materials. This means that there are few opportunities for standardization or mass production except in small components.

Tables 3, 4 and 5 give details of some 4,800 construction firms in Canada, of which approximately 3,600 are corporate entities, although it has been estimated that there are in fact about 80,000 construction firms in Canada, of which approximately 20,000 are corporate entities. Among the incorporated firms, equity financing is proportionately much lower than in manufacturing or wholesale trade, while short-term debt, especially loans, accounts for a larger share of current liabilities. The low requirement for equity financing no doubt explains in part the ease with which new construction firms are formed.

Another unique feature of the construction industry is the unparalleled diffusion of responsibility for the finished product, particularly in the division between design and construction. In most industries the designer and the engineering department work closely in/ the development of a new product to ensure that their efforts complement each other, but in the construction industry the designer, at least on a stipulated sum contract, does not know who the contractor will be until the design is complete. He therefore doesn't know of any special construction techniques which the successful contractor might be able to use and which might have been incorporated in the design. This is overcome to a certain extent by the management contract, but even with this approach there is still a separation between the designer and the subcontractors.

THE BUILDING

To this point, all the factors mentioned have been factors external to, and beyond the control of, the design consultant. They are like ice on a road. The driver of a car tries to make himself aware of its existence, he takes all the corrective actions when he finds himself on the ice, but if it goes out of control despite his precautions there isn't much he can do about it.

At the individual building level, on the other hand, the design consultant does, to a greater or lesser degree, have some scope to anticipate and control the factors affecting cost. These factors are of various kinds.

Type of Building

The first is the type of building. Once the building owner has decided what kind of building he wants to put up he has automatically selected a magnitude of construction cost. Each building type has its own range of construction costs and it is impossible, for example, for a hospital to have the same square metre cost as a parking garage. This may appear self-evident but there are a surprising number of people to whom this comes as a revelation.

The reasons why each building type has its own range of construction costs are that the type of use and occupancy will determine the general standard and quality of the building systems and finishes. Also the building type often governs the kind of contractor who will build it. Non-union contractors and sub-contractors are customarily associated with residential construction, union contractors and sub-contractors are usually associated with institutional buildings, while industrial and commercial buildings may be built by either union or non-union contractors and sub-contractors. There is a great difference in cost (to the building owner) between work performed by union and non-union contractors.

Location

The location of the building can have a considerable influence on the construction cost in a variety of ways. It is well known that the cost of labour, together with the permitted number of working hours and the fringe benefits, will vary from location to location, and so will the quality of the labour. Usually the most productive labour is to be found in the major urban areas where, not surprisingly, the labour rates are higher. However, on a unit of production basis variations in productivity and labour costs tend to cancel each other out, so unit prices are not quite so variable as might at first be supposed. A low productivity with a low labour cost gives a figure not dissimilar to a high productivity with a high labour cost. On the other hand, labour is more readily available in major urban areas so there is less likelihood of labour shortages and consequent delays.

Ontario and Quebec supply virtually all their own construction materials and are able to export materials to other provinces. The Atlantic Provinces have to import about 70% of their materials, the Prairies about 50% and British Columbia about 40%. Even within the central provinces there can be some variation. Materials at competitive prices are readily available in the large towns but are obtained with more difficulty and hence at higher cost in the rural areas, which means making use of local materials whenever possible. In Arctic areas materials become extremely expensive because they all have to be imported and transportation costs are very high. Great care must be taken in ordering materials in these remote areas because the omission of a few pounds of nails can delay a project until the next shipment of supplies arrives – there is no local hardware store to make good the deficiency.

All municipalities have planning bylaws governing the size and height of buildings but they vary from municipality to municipality and some are more restrictive than others. Since the size, shape and height of a building will have a direct bearing on its cost, then indirectly the municipal bylaws in operation at the building site will also have an effect.

When a building is being built as a speculative venture its location will determine the type of rents it can command, and this in turn will determine its general quality. A building on a prime site commanding high rents will be of a higher quality and will cost more to build than one on a poor site with low rentals.

A building on a downtown site in a large city may have very restricted access, requiring traffic control when materials are being delivered or requiring that materials can only be delivered at certain times of the day. There may also be difficulties in storing materials on the site because of space limitations. All these restrictions will add to the cost. On suburban or country sites there are usually no problems of access or storage although access can be a major problem in Arctic areas. In the Arctic access may only be feasible at certain times of the year and facilities such as roads, boat docks or aircraft landing strips may have to be provided to make it possible at all. Also, the restricted access may make it necessary to build canteen, recreation and health facilities in addition to sleeping accommodation for the workmen. It is not unknown for the cost of these temporary facilities to equal or exceed the cost of the building being constructed.

The location of the site will determine its cost, sites in major urban centres costing rather more than sites in rural areas. This in turn will govern the type and quality of the building which is constructed and possibly its size, shape and height.

The Site

Closely allied to the location of the site is the site itself. A building owner who is going to erect a building on a confined site with other buildings adjoining it, and possibly with a major street running beside it, will probably need to go to the expense of shoring the excavation, and may need to underpin some, if not all, of the adjacent buildings. This is a problem which can easily be overlooked when preparing a preliminary estimate on a project. A confined site can influence the shape and height of the building, while a large site puts no such restrictions on the building unless it is an undulating site, in which case the form of the building should take advantage of the contours.

The type of soil conditions found at the site will influence the design of the foundations, will determine whether special costs such as sheet piling or pumping will be required, and will govern the unit prices for excavation and other foundation work.

It can reasonably be said that the location and the condition of the site can have a major influence on the construction cost.

Shape

The shape of the building can also have a major effect on the construction cost. Figure 1 shows the plans of several buildings of various shapes and sizes, and Table 6 shows the data associated with them. Buildings A - E all have a gross floor area of 10 000 m² but while Building A is square, Building B is rectangular. This change in shape gives Building B a longer perimeter than Building A and if the exterior cladding is ten metres high in both cases this will result in a cladding area of 5 000 m² for Building B and 4 000 m² for Building A. At a cost of $100.00 per square metre for cladding, Building A will have a cladding cost which is $100 000 less than Building B, or a differential of $10.00 per m² of the gross floor area.

Compared to the simple square shape of Building A, Buildings C, D and E all have shapes which have increased the cost to varying degrees, and it is generally true to say that any variation from the square, other than to a circle, will increase the cost of cladding because the ratio of exterior wall to enclosed space is increased. These examples show the effect on the cladding cost, but the cost of air conditioning and heating will also be increased because of the increased cladding ratio. It is also worth noting that exterior corner columns, being eccentrically loaded, are less economical than interior columns, and this will also add to the cost of Buildings C, D and E.

The most efficient shape in terms of the ratio between its perimeter and its area is the circle, but circular buildings, while not unknown, are usually more expensive to set out and to build and also give the designer problems when it comes to planning the interior.

Table 6

Building	Area	Perimeter	Cladding Area	Cladding Ratio	Cladding Cost	Cost per m²
A	10 000 m²	400 m	4 000 m²	0.40	$ 400 000	$ 40.00
B	10 000 m²	500 m	5 000 m²	0.50	$ 500 000	$ 50.00
C	10 000 m²	600 m	6 000 m²	0.60	$ 600 000	$ 60.00
D	10 000 m²	420 m	4 200 m²	0.42	$ 420 000	$ 42.00
E	10 000 m²	530 m	5 300 m²	0.53	$ 530 000	$ 53.00
F	250 000 m²	2 000 m	20 000 m²	0.08	$ 2 000 000	$ 8.00

Size

Size has a significant effect on the cladding cost as can be seen by Building F in Table 6. This building is square, as is Building A, but its sides are five times longer which results in a ratio of cladding area to floor area of only 0.08, and a saving of $32.00 in the cost per square metre of the gross floor area. It is generally true that the larger a building is on plan the lower its cost per square metre will be. This means that it is normally less expensive to incorporate a given amount of floor space in one large building than it is to spread it out over a number of smaller buildings.

FACTORS INFLUENCING COST

FIGURE 1.

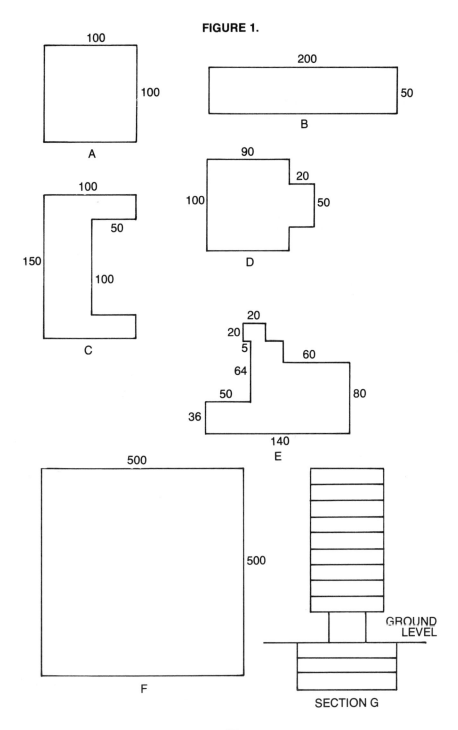

Height

It is less easy to generalize about the effect of height on cost than it is to generalize about shape and size. The height of a building is often governed by the site and the local planning bylaws, but for a given floor area it is generally, but not invariably, true to say that a single-storey building will cost more than a two-storey building. This is because the single-storey building has twice the area of foundations, slab on grade and roof finish, while the area of roof construction is equal to the combined areas of floor and roof construction in the two-storey building. Even the improved cladding ratio and the introduction of stairs does not usually offset these premiums. Exceptions to this rule are where floor loadings are large, necessitating an extra heavy construction for the upper floor in the two-storey building, or where the roof construction can be quite light making its unit cost considerably less than a suspended floor. Both these conditions usually prevail in warehouse and factory buildings and this, combined with the ease of material handling in a single-storey as compared with a multi-storey building, is why these types of buildings are normally single-storey structures.

Above two storeys each additional floor adds to the cost for a given floor area. Additional floors reduce the plan size and thereby make the cladding ratio less efficient. They also reduce the ratio of net to gross floor area because more elevators have to be introduced as the building becomes higher, besides the additional cost of the elevators themselves. As well as these considerations, electrical and mechanical systems become more complicated, the structure of the lower floors has to be designed to carry the upper floors, the building has to be able to withstand wind loadings, and construction costs will increase with multi-storey buildings because of hoisting equipment costs and the extra time needed to get men and materials to their locations in the building.

The other aspect of height – the storey height of each floor – is usually determined by the building owner's requirements and the dictates of the structural and mechanical designs. It is obvious that any increase in storey height will directly affect the ratio between the area of cladding and the gross floor area and increase cost.

Projections

While the shape of the building on plan has an effect on the construction cost, the shape on section can also have an effect. Projecting or recessed balconies require additional expense for their construction and finish, and the latter will change the shape of the building on plan and increase its cladding ratio.

Buildings which have a ground, or any, floor which is smaller than the floors above and below it, such as Building G shown in Figure 1, will have additional costs. The soffit of the second floor where it is exposed will have to be insulated and finished, the roof over the basement immediately under the second-floor soffit has to be waterproofed and paved whereas normally it would form part of the ground floor, and the columns supporting the perimeter of the second floor will probably require special treatment. The net effect of this design is to reduce the total floor area while at the same time increasing the total cost.

Exterior Cladding

Because exterior cladding seems to crop up so often in any discussion of the effect of the size, shape and height on the cost of a building, it might be worthwhile examining why it appears to be so important. One of the reasons is that it is the largest element which can be manipulated by the architect and the building owner. With many of the other elements in the building there are strict limits governing the variations which the architect and his engineers can make in order to affect their costs, but by changing the size, shape or height of the building they can readily change the quantity of cladding and hence its total cost. It is therefore rather more flexible than most of the other elements.

In comparison with interior partitions, exterior cladding is expensive because it needs to be weatherproofed and insulated, and it only encloses space on one side. For these reasons it is preferable, if possible, to reduce the quantity of exterior cladding even though it means increasing the quantity of interior partitions.

In addition to the cost of the cladding itself, the exterior cladding is where heat losses and heat gains are made, so the quantity of cladding, including the roof, has a great effect on the capital and running costs of the heating and cooling systems.

Planning Efficiency

Before the final shape and height of a building is settled a careful look at the planning of the interior spaces is needed. A building which has a good shape in terms of the ratio of exterior cladding to enclosed space may require an increased amount of circulation area, or make it difficult to incorporate storage or mechanical spaces within it. A saving of five per cent in the cost per square metre because of a good cladding ratio is wasted if it has caused the building to be enlarged to a gross floor area which is ten per cent more than it needs to be. The question of net to gross floor areas should always be studied at an early stage of the design process since it is not unknown for a building to appear to be economical on the basis of its square metre cost, while in fact the building owner is buying more building than is really needed.

Materials

Considerations of size, shape and height relate to decisions which are made in the early stages of a project and once these decisions have been made the cost implications have largely been settled. It then remains for decisions to be made as to the type and quality of materials and systems which, while important, do not have the same major cost effects.

The prime requirement for economy when it comes to working details is simplicity. Uncomplicated, clear and unambiguous drawings and specifications not only help reduce costs on the site but also ensure that truly competitive tenders are received. A contractor's estimator who doesn't really understand the drawings or specifications is likely to cover himself with an inflated price, and the building owner will be paying extra for his designer's obscurity.

Allied to simplicity is the designer's need to recognize the stock sizes of materials so that they can be incorporated into the building with a minimum amount of wastage, and to use standard manufactured components rather than custom-made products. Using expensive materials where less expensive materials would be quite adequate not only increases the cost of materials unnecessarily but also increases the cost of labour. A walnut door, for example, takes more time and trouble to hang just because it is an expensive door, than does a simple hollow-core birch door.

Finally, formwork is the most expensive single element in concrete construction. When the costs of a reinforced concrete structure are analyzed it is usually found that the formwork costs far exceed the cost of the concrete or reinforcing steel. This is the reason for some structural designers maintaining the same size of columns all the way up the building, making the necessary adjustments in the amount of reinforcing steel, in order that the column forms can be re-used without having to be re-worked on each floor. In fact any saving in formwork, such as the use of flying forms, will help reduce the costs of concrete construction.

Type of Contract

The type of construction contract which is entered into can have some influence on costs. The stipulated sum contract, in which all contract documents are prepared in advance of construction and a group of general contractors submit lump sum tenders based on those documents, is well known. This type of contract has several advantages. It is well understood so there are few opportunities for misinterpretations. It is completely competitive so, in theory at least, the building owner pays the lowest possible price and the building owner knows, subject to changes during construction, what he will have to pay for his building before he is committed to it. Its disadvantages are that the design team works in complete isolation from the contractor and cannot take advantage of any innovative methods he might be able to suggest. Only the contractor knows how he established his price and a lump sum tender doesn't give the design team much useful cost information for future projects. Also the time required for preparing complete working drawings and specifications can be very prolonged, delaying the owner's occupancy of a completed building by several months.

Another familiar form of contract is the cost plus contract in which the contractor agrees to build at his cost, plus a markup for his overhead and profit, usually stated as a lump sum or as a percentage of the cost. The only advantage of this method is that it enables the contractor to start work on the site much sooner than he would if he were operating under a stipulated sum contract since he doesn't have to wait until all the drawings and specifications are complete. It might also be used for a complicated renovation project where it is impossible to estimate the cost in advance. Its disadvantages are numerous. The building owner doesn't know what his cost commitment will be until the work is finished and the final bill presented. When the markup is based on a percentage of the cost the contractor has a positive incentive to increase the cost and will usually put his least experienced supervisors and workmen on the job. And there is often very little competition, particularly on smaller projects when a single sub-contractor may be asked to submit a price with no other sub-contractor being asked to quote against him. In comparison with a stipulated sum contract, therefore, the cost to the building owner is almost bound to be more if he enters into a cost plus contract.

A cost plus contract with a maximum upset price at least ensures that the building owner knows what his maximum cost commitment will be. The other disadvantages of the cost plus contract remain, however, and the wise contractor will make sure that the maximum upset price is sufficiently high to cause him no problems. A participation clause which allows the building owner and the contractor to share any savings between the actual cost and the maximum upset price on an agreed ratio gives the contractor some incentive to reduce costs, but he still has the incentive to make his maximum upset price as high as he feels he can. Problems can also be encountered in agreeing upon what constitutes a saving in which both parties participate. This type of contract, like the ordinary cost plus contract, is likely to be more expensive to the building owner than a stipulated sum contract.

A management contract, also known as fast-track, is not unlike a cost plus contract although there are some crucial differences. A construction manager, who may be an individual, a construction management firm or a construction company is appointed to manage the construction. The appointment may be made by selecting a firm or an individual in whom the building owner and the design consultants have confidence, and with whom the building owner can negotiate a fee, or by competitive tender among a group of firms or individuals. If the latter method is used it should be remembered that the essence of a management contract is good construction management. This is not always provided, and quite possibly will not be provided, by an individual or firm submitting the lowest fee. Once the construction manager has been appointed he works with the design team during the development of working drawings and specifications, and manages the work of construction on the site much as a general contractor would on a stipulated sum contract. However, drawings and specifications are prepared in sequence in trade packages and tenders are called from the sub-trades as they are required to meet the construction schedule. Thus the architect will, for example, concentrate on designing the roofing when the next trade package in the construction sequence requires a roofing contract, and will only concentrate on finalizing finishes when they are the next in sequence.

The advantages of a management contract are that, like the cost plus contract, work can start on the site with very little delay and before all the drawings and specifications are finalized. The building owner and his consultants are privy to all the costs which make up the total cost of the building; and, provided the management contractor is not required to do any of the construction work, all prices are competitive since the sub-trade packages are called by competitive tender. Another, sometimes illusory, advantage is that the construction manager is available to give practical advice to the design team almost from the beginning of the project. He cannot usually be brought in right at the beginning of the project because he needs to have some knowledge of the type, size and standard of quality of the building before he quotes his fee.

The main disadvantage of this type of contract is that the building owner doesn't know what his cost commitment is until the last tender has been called and it is theoretically possible for the building owner to run out of money halfway through construction if the early trade packages exceed the budgets by substantial amounts. The actual construction cost to the building owner is likely to be about the same as it would be with a stipulated sum contract, his principal saving being in the earlier occupancy of the building, but the costs to the design team are likely to be greater since they are put to considerably more work.

Design Consultants

The architect or his engineers can have an effect on construction costs. Some consultants have a reputation for being expensive, based not just on the way in which they design buildings but also on such factors as the quality of their drawings and specifications and their attitude toward the contractor and his sub-contractors. A contractor soon learns which architects and engineers will be unreasonable in their demands during construction, or will delay him by not producing the information he needs and, accordingly, will add up to ten per cent to the tender price to cover the additional costs entailed.

CHAPTER 3
Estimating Methods

THE CONTRACTORS' METHOD

When a contractor submits a tender he does so on the basis of a detailed breakdown of all the quantities of labour and materials which will be required for the project, to each of which he applies a unit price which will enable him to estimate the total cost. This method is usually referred to as the quantity survey method. At this stage, it should be remembered, it is only an estimate of the total cost since it is only when the building is actually finished that the contractor knows the real cost, and whether or not he has made a profit; although a prudent contractor will have a good idea of how his estimate compares with the actual costs right from the start of construction.

The measurement of quantities, the first stage in the process, is a technical function requiring a good knowledge of construction and an ability to visualize the work as it will be built. Since there are a number of ways in which the quantities can be measured and presented for pricing, the Canadian Institute of Quantity Surveyors has published a *Method of Measurement of Construction Works* which sets out the general rules to be followed to provide a degree of standardization in the measurement and presentation of the quantities. Many contractors follow this document and it is used as a basis for teaching the measurement of quantities in most technical colleges.

The second stage in the process, the application of unit prices, requires the same skills as those needed for measuring the quantities, together with a detailed knowledge of site operations.

There are three basic elements of a unit price: labour, material and equipment. Labour is undoubtedly the most difficult to estimate, requiring a detailed knowledge of the time required to perform all the various operations on the site, and an understanding of what factors can affect them. Most contractors have some form of reporting system from the site to the head office. In its simplest form it may only be a weekly report of the total cost of labour for payroll purposes. Without any breakdown or explanation of the work which was being performed, this information is not of much help to the estimator.

A more useful form of report would give a breakdown, either in man-hours or in labour cost, of each of the operations being performed on the site. This enables head office to compare the actual costs with the estimated costs, and the estimating department can record the information for future estimates. Care has to be taken in analyzing this material because a busy site staff can distort reports unintentionally if it hasn't had time to prepare the information properly, while an unscrupulous superintendent can distort reports deliberately to cover up errors.

If the report is based on the labour costs rather than on man-hours, the estimating department would have to analyze the costs back into man-hours, to give labour constants. A labour constant is a representation of the number of man-hours required to perform a given operation. The labour constant for hanging a 9800 mm x 2000 mm hollow-core, birch-faced flush door, for example, might be 0.8 hours carpenter time plus 0.4 hours labourer time. These times, multiplied by the appropriate labour rates, will give the total cost of hanging a door, as follows:

0.8 hours carpenter @ $9.64 = $ 7.71
0.4 hours labourer @ $7.66 = $ 3.06
Total $10.77

While these labour constants may have been obtained from one project, the next project may show a different set of constants and it is up to the estimator to find out, if possible, why they differ. Once the estimator has discovered the differing conditions which occurred for the same operation on two different projects he is in a better position to allow for them on future estimates.

An alternative to the labour constant method of estimating costs, and one which is perhaps a little more practical, is the crew cost method. It also requires feedback from the site, but instead of analyzing the information into labour constants it is analyzed to show the production which might be expected from a crew of workmen in a day. Using the example of hanging a door again, the crew may consist of two carpenters and a labourer, and their production might be twenty doors in a day. The calculation is therefore:

2 carpenters x 8 hours = 16 hours @ $9.64 = $154.24
1 labourer x 8 hours = 8 hours @ $7.66 = 61.28
$215.52

The unit cost per door will be $\frac{\$215.52}{20} = \$10.78$ each

This method has certain advantages over the labour constant method. It is easier for the site staff, if they have occasion to review the estimate, to visualize a crew of men producing a certain volume of work in a day than it is for them to visualize the rather abstract labour constants, and for the same reason it is easier for the estimator to adjust for site conditions and to make allowances for changes in productivity when he is preparing his estimate. Also the estimate can more easily be used as a means for determining site labour requirements when the contractor is scheduling his job.

It should be noted that the labour rates used in the calculations above are the hourly rates to be paid by the contractor and not the hourly rates paid to the workmen. Such extras as holidays with pay, workmen's compensation, union checkoffs and other fringe benefits that are paid by the employer over and above the workmen's hourly rate have to be included in the estimate. These can be added to the hourly rate as has been done here, or they can be carried as a percentage on the total cost of labour at the end of the estimate as is done by some contractors. However it is done, these extras have to be included in the estimate, otherwise their cost will come out of the contractor's anticipated profit.

Material is easier to estimate than labour because it is usually just a matter of obtaining quotations from suppliers. But the contractor's estimator must check whether the quoted price includes the cost of delivery to the site or whether he has to add this cost to the price, whether discounts are available, whether sales taxes have been included and whether duty has to be paid on imported materials.

A major problem with estimating some material costs is determining what allowance should be made for waste. The *Method of Measurement of Construction Works* states that all measurements shall be made net, which means that the unit prices have to allow for all the wasted materials. Many materials have waste associated with them: plywood panelling which is bought in 1200 mm x 2400 mm sheets will have at least one-eighth wasted if it is to be installed on walls which are 2100 mm high, wood framing always has to be cut to correct lengths with a consequent waste of material, all the bricks and blocks which have been purchased don't finish up in the building and a good proportion of the mortar usually ends up on the ground. Waste is usually allowed for as a percentage addition to the material cost, the actual amount varying from material to material, but the extent of wastage will also depend partly on the detail design of the building and partly on the calibre of the workmen. If inefficient workmen waste more materials than was allowed for in the estimate, the contractor's profit will be less than he had anticipated.

If the *Method of Measurement of Construction Works* is not being used as a guideline, waste allowances can be incorporated in the measurement by adding the appropriate percentages to the quantities, in which case no allowance needs to be made in the unit prices.

The cost of equipment can either be included as part of the unit price, or it can be added at the end of the estimate as one of the items of site overhead.

The cost of vibrators for concrete, for example, is usually included as part of the cost of placing the concrete, while the cost of the crane used for hoisting the concrete into place is usually included in the site overhead because it is used for hoisting other materials besides concrete. The cost of the equipment is calculated on the time it will be required on the job. Added to this figure is the cost of transporting the equipment to and from the site, erecting and dismantling it, the cost of labour operating it and the fuel and other supplies needed to run it. Some equipment is rented, but even if it is owned by the contractor he will charge it to the job at current rental rates so as to cover his ownership costs.

When the contractor has estimated the direct cost of labour, materials and equipment for the work which will be done by his own forces on the project he adds the cost of work which will be sub-contracted, any cash allowances which have been included in the specification, the cost of site overheads, and his head office overhead and profit. Site overheads, besides major equipment which has already been mentioned, will include supervision, insurance, bonds, hoarding, site offices and sheds, permits and so on. A summary of items usually included as site overheads is given in Appendix B under Element 10. Head office overhead and profit, unlike the rest of the estimate which can be estimated directly, is calculated as a percentage of the total cost or is included as a lump sum. It is intended to cover the cost of the contractor's office staff; heat, light and rent of the office; office supplies; and an allowance for profit.

A contractor's estimator, if the contracting firm is at all efficient, will have most of this detailed cost information readily available. For those who are not working in a contractor's office the following publications may be of some help in obtaining cost information:

Building Construction Cost Data, published annually by Robert Snow Means Company Inc., 100 Construction Plaza, Duxbury, Mass. 02332.

This is a compendium of unit prices based on average costs throughout the United States. It is very comprehensive and can be useful when other sources are not readily available.

Construction Pricing and Scheduling Manual published annually by McGraw-Hill Information Systems Company, 330 West 42 Street, New York, N.Y. 10035

This publication is similar to Means.

Yardsticks for Costing, published annually by Southam Business Publications Limited, 1450 Don Mills Road, Don Mills, Ontario.

This publication, while not as comprehensive as those produced by Means or McGraw-Hill, contains unit prices for seven cities in Canada and has a quarterly updating service.

Boeckh Building Valuation Manual (3 volumes), published by General Appraisal of Canada Limited, Publication and Education Division, 170 University Avenue, Toronto.

These books give costs for a large variety of buildings together with unit prices for a number of different materials and are supplemented by a bimonthly publication giving updating factors for the costs. The whole enterprise is directed primarily toward appraisers.

The Building Estimator's Reference Book, published by Frank R. Walker Company in Chicago.

This book, which can be bought in any good technical bookshop, does not contain unit prices but it does have labour constants and other estimating data which can be useful in establishing unit prices.

Estimating Construction Costs (third edition), by Robert L. Peurifoy, published by McGraw-Hill Book Company Inc.

This book is similar to Walker's book but is not quite as comprehensive.

The first four publications listed can be quite useful, but should be used with some discretion since they contain average prices which may not apply specifically to any one particular project. They can therefore be of use to an architect, engineer or quantity surveyor who is trying to establish a budget, or who is making a comparison between different types of construction or finishes, but they would never be used by a contractor who is submitting a tender.

Other ways of obtaining cost information for those not employed with construction companies include personal observation on the site, negotiating the cost of changes with contractors, and talking to contractors, sub-contractors or suppliers to find out current unit prices. This last has to be done warily, particularly in the case of specialist sub-contractors and suppliers. Some tend to quote high figures so that there will be no recriminations afterwards, while others tend to quote low figures in the hope that it will ensure that their product is specified. A careful assessment must be made to verify the accuracy of the figures.

THE DESIGN CONSULTANT'S METHODS

The only way a contractor can prepare his estimate is to make a detailed analysis of all the labour, material and equipment which will be required for the job, with a trade-by-trade breakdown of the costs. This method is not usually appropriate for the design consultant for the following reasons:

1. The contractor works out his estimate from detailed drawings and specifications; but the design consultant, particularly in the early stages, has less information on which to base his estimate and couldn't measure detailed quantities even if he wanted to.

2. The contractor knows that, provided there are no unforeseen delays, if he is low bidder he will be able to start construction shortly after completing his estimate and can assess accurately the likely duration of the project. The design consultant, on the other hand, may be preparing a preliminary estimate on a project which may not start construction for two years or more, and which may have a rather indeterminate duration because of the lack of detailed design information. As a result he is trying to project costs much further into the future than the contractor, and although he may include a contingency to cover cost escalation the measurement of detailed quantities with what may turn out to be inaccurate unit prices would be pointless.

3. The contractor has an estimating department and a substantial amount of detailed cost information flowing in from the site, together with firm price quotations from sub-contractors and suppliers. Most design consultants do not have an estimating department, and very often the only detailed cost information they have is a trade breakdown from a previous project. If the previous project was a stipulated sum contract the breakdown is likely to be quite misleading because it was provided by the contractor for progress payment purposes, not to help the design consultant with his estimates. It will therefore be distorted by having money transferred from the later trades into the early trades so as to help the contractor finance the project. Given that he is functioning without an estimating department and with inadequate cost information the average design consultant is usually not able to make an estimate in the same amount of detail as a contractor.

4. The contractor prepares his estimate with the object of obtaining work. It has to be as accurate as is humanly possible, otherwise, despite being low bidder and being selected to do the work, he may lose money on the project. The design consultant's estimate is prepared to advise the client of his financial commitment and to help control costs. It therefore does not have to have the same degree of accuracy. Since the objectives are different, the estimating methods can also be different.

5. The contractor needs to break his costs down into a trade format because this is how his own cost information is organized and because this is how he receives his sub-trade prices. As will be explained later, a trade breakdown is of little help to the design consultant in controlling costs in the pre-tender period.

6. The contractor's method of estimating is very accurate but extremely time consuming. The design consultant, who may have to prepare several estimates on alternative design solutions, needs a method which is, perhaps, rather less accurate but much quicker.

Accepting the fact that the contractor's method of estimating is not generally suited to the design consultant, those methods which he can use should be looked at with three requirements in mind. First, they should be reasonably accurate since there is no point in using a method which is notoriously inaccurate. Second, they should be reasonably fast because they are used for making decisions, and decisions usually have to be made quickly. Third, they should provide a means of controlling as well as estimating the cost. The first two requirements are always in conflict with each other; a fast estimate is usually inaccurate and an accurate estimate usually requires time for its preparation, but a reasonable balance must be reached between the two.

All estimating methods can be classified as single-rate or multiple-rate methods, and traditionally design consultants have tended towards single-rate methods.

Unit Cost Method

The first single-rate method is the unit cost method in which the cost is analyzed as a cost per unit of accommodation. Examples would be the cost per car in a parking garage, the cost per student in a school, the cost per bed in a hospital and the cost per suite in an apartment building. The cost is analyzed by the simple division of the known total cost of a building by the unit of accommodation, and an estimate is made just as simply by multiplying the known number of units in a proposed building by the previously analyzed cost per unit. This method is very fast but not very accurate because it is difficult to make adjustment to the unit cost for changes in size, shape, quality or type of construction. Nor is it useful in controlling costs because it doesn't give enough information about what is included in the costs. While it is not recommended as an estimating method, it can be used by a school board or a hospital commission for establishing overall construction budgets and it can be helpful as a check on other estimating methods. It is also useful as a means of comparison of known costs between one building and another, particularly in the case of parking garages or apartment buildings.

Square Metre Method

The square metre method (formerly the square foot method) is probably the most common method used by design consultants, and it is certainly the most common way of discussing building costs. It is also the method which has caused many an estimate to go wrong.

The principal advantage of this method is that the gross floor area is related to usable space and is therefore readily understood by the building owner, and, in theory at least, it should be reasonably accurate because a major part of the cost of a building is contained in, or related to, its horizontal components. Like the unit cost method the cost is analyzed by simple division, and the estimate is made by multiplying the gross floor area of a proposed building by the previously analyzed cost per square metre.

Because of its common use as both an estimating method and as a means of comparing construction costs it is important that the terms used in the square metre method be adequately defined. The Canadian Institute of Quantity Surveyors' booklet *Measurement of Buildings by Area and Volume* sets out the definition of the measurement of the gross floor area, which is now generally accepted. (See Appendix A.) At the same time, the cost should also be defined, although there is no generally accepted way of doing so. The cost is always recognized as the cost which the contractor charges the building owner for a building, not the cost to the contractor for building it. However, the cost of one building may, for example, include the furnishings and site works while another excludes them. A discussion of square metre costs in which the parties each have a different interpretation of the meaning of the total cost of a building, even though they have both measured their buildings in the same way, becomes meaningless.

This method, like the unit cost method, is quite fast but not particularly accurate because again it is not easy to make adjustments to the unit price for changes in size, shape, quality or type of construction. It is also not helpful in controlling costs. Its main use might be in checking results determined by some other estimating method.

Cubic Metre Method

The cubic metre method is a variation of the square metre method, the measurement going one step further and incorporating the height as well as the area to give a cubic metre quantity. At one time it was a very popular method, but it is rarely used now, because it has two major defects. The first of these is that, while it takes the third dimension into account, doing so tends to produce misleading information. The measurement of the volume of a building doesn't give any indication of what is contained within that volume. A cubic metre cost may be high either because the designer has been able to incorporate more usable space into the building, perhaps by reducing the storey heights in order to introduce an additional floor into the same volume, or because he has used more expensive materials or construction details. It is therefore difficult to appreciate why two buildings have different cubic metre costs; the reason one is higher than the other may be either because the designer is more efficient or because he is more expensive, and it is not immediately apparent which answer applies.

The other major defect of the cubic metre method is that a large volume is multiplied by a small unit price, and a small variation in the unit price will result in a large variation in the total cost.

The cubic metre method also has the same drawbacks as the square metre method and these, combined with its two major defects, do not recommend it as a method of estimating cost.

It is evident that the three single-rate methods, the unit cost, square metre, and cubic metre methods, all use one quantity to be multiplied by one unit price – hence the name single-rate method. The one advantage of any single-rate approach is that the unit price incorporates all the costs to be found in the building so there is no danger of leaving anything out. Against this is the bigger disadvantage of an entire estimate being wrong if there is an error in the unit price. Using a single-rate method requires a great deal of experience in selecting the unit price if estimates are to be tolerably accurate.

Multiple-rate methods, on the other hand, carry the possible danger of omitting some items of work but against this an error in any of the unit prices does not necessarily invalidate the whole estimate, particularly since the law of compensating errors is likely to come into effect. Multiple-rate methods take longer to calculate than single-rate methods but are generally more accurate and they can be used as an aid in controlling costs. The most familiar multiple-rate method is the quantity survey method used by contractors which has been described in the opening section of this chapter.

Unit In Place Method

The unit in place method, not to be confused with the unit cost method, and sometimes known as the approximate quantities method, is a simplified form of the quantity survey method. With the quantity survey method, for example, a reinforced concrete wall would be estimated by measuring the cubic metres of concrete, the square metres of formwork and the mass of reinforcing steel, each listed and priced separately, and possibly with separate costs for labour, material and equipment. Applying the unit in place method the wall would be measured in square metres and a single portmanteau unit price used to cover the combined costs of labour, material and equipment for the concrete, formwork and reinforcing steel. Besides measuring walls this way, any number of components in a building can be treated in a similar fashion. A distinct benefit of this approach is that it is easy to make adjustments in the cost when, say, it is decided to make a change from a reinforced concrete to a block wall. No adjustment is needed to the quantity, only a change from one unit price to another. Whenever possible the measurement is simplified and the unit prices are adjusted to cover all the small items normally measured in great detail in the quantity survey method.

The main advantages of the unit in place method are that it is quite accurate, it takes into account the peculiarities of the building such as its size, shape, quality and type of construction, and because it identifies the costs of systems and materials in the building it can be of great help in controlling cost in the later stages of design. It is also the only method, other than the quantity survey method, which can be used for estimating the cost of alteration work.

Its disadvantages are that it takes considerably more time than any single-rate method, and it needs a certain amount of design information before it can be used. It cannot be used, for instance, at the sketch drawing stage, except for some elements, because there is insufficient design information available at that point to measure unit in place quantities. In fact, any attempt to measure unit in place quantities at that stage is likely to result in a great many items being overlooked, leading to a hopelessly inaccurate estimate.

Typical Bay Method

In the typical bay method, a typical bay is measured and priced in detail using the quantity survey method or the unit in place method, and the cost of the bay is multiplied by the total number of bays in the building to give a total cost for the building. As an estimating method its major defect is that it is not usually easy to identify a typical bay, and so many adjustments have to be made for non-typical bays that the estimate can get out of control. It is, however, useful for comparing the costs of different structural and cladding systems, and an analysis of several alternative designs based on typical bays will indicate which are the more economical, which might be considered further and which should be discarded immediately.

This method is not therefore recommended as a total estimating method, but because it is quite fast and accurate as a means of comparing costs it can be useful at the design development stage in helping to make design decisions.

Elemental Method

The final multiple-rate method is the elemental method which will be discussed in greater detail in the next chapter.

CHAPTER 4
Elemental Cost Analysis

Cost Analysis

No matter what type of estimating method is used, an analysis of known construction costs is needed to provide the necessary information for future estimates. Furthermore, the estimating method to be used will automatically determine the way in which costs have to be analyzed. If it is proposed to use the square metre method of estimating, then the analyses must be recorded as costs per square metre, and there is little point in recording them in any other form. There is therefore a close affinity between a cost analysis and the estimating method for which it will be used.

In fact, the only difference between the analysis of any of the single-rate methods and the estimate produced from the analysis is the difference between division and multiplication. The known cost of a building is divided by its gross floor area to arrive at its cost per square metre (the analysis), while the gross floor area of a proposed building is multiplied by a known cost per square metre to arrive at a total cost (the estimate).

In the case of the multiple-rate methods, while the same principle applies, instead of only one analysis being performed some form of breakdown is required and an analysis is performed on each of the items in the breakdown. In the quantity survey method, for example, one of the items to be analyzed might be the cost of labour to lay 150 mm concrete blocks, in which case the known cost of laying a certain number of blocks is divided by the number of blocks to give the labour cost per block. As each item is analyzed in this way a library of cost analyses can be built up and used for future estimates, bearing in mind of course that cost analyses can soon become out of date and have to be frequently revised.

Elemental Cost Analysis

The elemental method of estimating, being a multiple-rate method, requires a breakdown, and elemental cost analysis will require an identical breakdown. The methods described in the last chapter were either too simple or too detailed, and any new method, if it is to be an improvement over the older methods, should have a breakdown which is not as detailed as those used in the quantity survey or unit in place methods, yet it should not be so limited as to make the method little better than one of the single-rate methods.

A breakdown which might occur immediately to a contractor is a trade breakdown since he is used to thinking about building costs in terms of trades. If he were to analyze and record the total cost of each trade for all the buildings he tenders it is conceivable that he might build up a useful library of cost information. But the trouble with a breakdown of costs based on trades is that the first use of a cost analysis is usually to provide a budget estimate in the very early stages of design, a time when the design consultants have given little thought to the actual materials which will be used in the building. An analysis based on a trade breakdown is therefore not particularly useful in preparing a budget estimate because at this stage the trades cannot easily be identified.

An alternative to a trade breakdown is a breakdown of costs by systems or elements. An architect thinks of the elements of a building in terms of the function each will perform rather than in terms of what it is made of, at least in the early stages of design. The primary function of an exterior wall is to exclude the weather from the building, with possible secondary functions of supporting the floors and roof and providing a pleasing external appearance. The primary function is always the same, regardless of what the wall is made of, and provided it can perform its primary function satisfactorily it is usually only when the secondary functions come to be considered that it is necessary to decide whether it should be made of brick, stone, concrete or some other material. A breakdown which identifies the cost of providing an exterior wall around the building is therefore of much more use to the architect than a breakdown which identifies, say, the cost of masonry, since masonry can be used in a variety of ways in the building. A more meaningful question for an architect to ask is "How much did it cost to enclose the building with exterior walls?" rather than "How much did masonry cost?"

An element has been defined as "a major component common to most buildings which usually fulfills the same function or functions irrespective of its construction or specification." In selecting a list of elements as a breakdown for use in a cost analysis the primary aim therefore is ensuring that each element performs a readily identifiable function within the building. Coupled with this, each element should have a significant cost, and wherever possible each should be capable of measurement. The Canadian Institute of Quantity Surveyors has established a list of elements which fulfill these requirements and which it recommends that its member use.

They are:
1. Substructure
    (a) Normal Foundations
    (b) Basement

      (c) Special Foundations
- 2. Structure
    - (a) Lowest Floor Construction
    - (b) Upper Floor Construction
    - (c) Roof Construction
- 3. Exterior Cladding
    - (a) Roof Finish
    - (b) Walls Below Ground Floor
    - (c) Walls Above Ground Floor
    - (d) Windows
    - (e) Exterior Doors and Screens
    - (f) Balconies and Projections
- 4. Interior Partitions and Doors
    - (a) Permanent Partitions and Doors
    - (b) Movable Partitions and Doors
    - (c) Glazed Partitions and Doors
- 5. Vertical Movement
    - (a) Stairs
    - (b) Elevators and Escalators
- 6. Interior Finishes
    - (a) Floor Finishes
    - (b) Ceiling Finishes
    - (c) Wall Finishes
- 7. Fittings and Equipment
    - (a) Fittings and Fixtures
    - (b) Equipment
- 8. Services
    - (a) Electrical
    - (b) Plumbing and Drains
    - (c) Heating, Ventilation and Air Conditioning
- 9. Site Development
    - (a) General
    - (b) Services
    - (c) Alterations
    - (d) Demolition
- 10. Overhead and Profit
    - (a) Site Overhead
    - (b) Head Office Overhead and Profit
- 11. Contingencies
    - (a) Design Contingency
    - (b) Escalation Contingency
    - (c) Post-Contract Contingency

A detailed description of these elements, prepared by the Canadian Institute of Quantity Surveyors and published with their permission, including a description of how they are measured for preliminary estimates is included in Appendix B.

This list of elements differs somewhat from those used in the United Kingdom where elemental cost analysis and cost control was originally developed, and any list is bound to be subject to some criticism from individuals who believe it could be improved. For example, it has been maintained that WALLS BELOW GROUND FLOOR are in fact part of the SUB-STRUCTURE and should be included in Element 1. This ignores the primary function of WALLS BELOW GROUND FLOOR, which is not to support the building as the foundations do, but to enclose the building as the WALLS ABOVE GROUND FLOOR do. The WALLS ABOVE GROUND FLOOR happen to keep out the wind and weather while the WALLS BELOW GROUND FLOOR keep out the surrounding soil, but the function of each is the same. A more serious criticism is that concrete partitions are frequently used to support the floors and roof, and their function is more structural than it is a dividing partition. While this is perfectly true, to maintain consistency these partitions should still be included under PERMANENT PARTITIONS AND DOORS and not in Element 2, with perhaps a note of what they cost attached to the analysis so that these figures can be extracted at a later date if the analysis is used to compare the cost of structural frames. Other criticisms of the present list include the comment that trades should be identified within the elements, which shows a misunderstanding of the use of an elemental cost analysis. Another criticism is that insufficient breakdown is allowed for SERVICES, which perhaps has some merit. Despite these complaints Canada does at least have one list of elements which is generally recognized by the majority of those working in the field of construction cost control.

Having decided upon a form of breakdown, the next step is to provide a means of expressing the results. A lump sum amount against each of the elements in the breakdown might be of some use if every building were the same size and shape. But since no two buildings are alike the known cost of each element has to be analyzed to put it into a useable form. This might be done by calculating the percentage each element represents of the total cost of the building. This is frequently done with mechanical and electrical costs, usually as a check to see whether they look adequate when putting together a pre-tender estimate, but percentages can be misleading. When checking mechanical and electrical estimates, for instance, there is sometimes a doubt about whether the mechanical and electrical costs appear low because a disproportionately high amount of money has been allocated to the rest of the building, or because they have been underestimated. This will apply with even more force if all the elements are calculated as percentages since a particularly high (or low) percentage in any one element will distort all the other elements and make comparisons between one building and another difficult.

A more suitable way of analyzing the cost is to show the cost of each

ELEMENTAL COST ANALYSIS

Helyar & Associates
Helyar, Rae, Mauchan & Hall Limited
Chartered Quantity Surveyors-Construction Consultants

Project University Student Residence Architect Tracey Prior Associates
Gross Floor Area 1696 m²

Project No. 78102
Date March 1978
Estimate No. Analysis

ELEMENT	RATIO	QUANTITY	UNIT RATE	AMOUNT	COST PER m² GROSS	TOTAL
1. SUBSTRUCTURE						
a) Normal Foundations				7700	4.54	
b) Basement				—	—	
c) Special Foundations				—	—	
Caissons etc. $						
Underpinning $						
Total					4.54	7700
2. STRUCTURE						
a) Lowest Floor Construction				8800	5.19	
b) Upper Floor Construction				66700	39.33	
c) Roof Construction				36500	21.52	
Total					66.04	112000
3. EXTERIOR CLADDING						
a) Roof Finish	—			12800	7.55	
b) Walls Below Ground Floor				—	—	
c) Walls Above Ground Floor				49500	29.19	
d) Windows				40700	24.00	
e) Exterior Doors and Screens				3500	2.06	
f) Balconies and Projections				800	0.47	
Total					63.27	107300
4. INTERIOR PARTITIONS AND DOORS						
a) Permanent Partitions and Doors				44900	26.47	
b) Movable Partitions and Doors				3300	1.95	
c) Glazed Partitions and Doors				—	—	
Total					28.42	48200
				CARRIED FORWARD	$162.26	$275200

FIGURE 2.

CONSTRUCTION ESTIMATING AND COSTING

Helyar & Associates

Page 2

ELEMENT	RATIO	QUANTITY	UNIT RATE	AMOUNT	COST PER m²	GROSS	TOTAL
				BROUGHT FORWARD		162.26	275 200
5. VERTICAL MOVEMENT							
a) Stairs				8 700	5.13		
b) Elevators and Escalators				—	—		
Total						5.13	8 700
6. INTERIOR FINISHES							
a) Floor Finishes				51 600	30.42		
b) Ceiling Finishes				22 400	13.21		
c) Wall Finishes (Wall Ratio ———)				82 300	48.53		
Total						92.16	156 300
7. FITTINGS AND EQUIPMENT							
a) Fittings and Fixtures				17 700	10.44		
b) Equipment				1 800	1.06		
Total						11.50	19 500
8. SERVICES							
a) Electrical				51 200	30.19		
b) Plumbing and Drains				33 200	19.57		
c) Heating, Ventilation and Air Conditioning				84 900	50.06		
Total						99.92	169 300
9. SITE DEVELOPMENT							
a) General				16 400	9.67		
b) Services				6 800	4.01		
c) Alterations				—	—		
d) Demolition				—	—		
Total						13.67	23 200
				CARRIED FORWARD		$334.55	$652 200

FIGURE 2. (continued)

ELEMENTAL COST ANALYSIS

Helyar & Associates

Page 3

ELEMENT	RATIO	QUANTITY	UNIT RATE	AMOUNT	COST PER m² GROSS	TOTAL
				BROUGHT FORWARD	384.55	652 200
10. OVERHEAD AND PROFIT						
a) Site Overhead				56 000	33.02	
b) Head Office Overhead and Profit				13 000	7.67	
Total					40.68	69 000
				SUB-TOTAL	425.24	721 200
11. CONTINGENCIES						
a) Design Contingency						
b) Escalation Contingency						
c) Post-Contract Contingency						
Total						
				TOTAL	$ 425.24	$ 721 200

COMMENTS:

FIGURE 2. (continued)

element as the cost per square metre of the gross floor area. The analysis is then not unlike the square metre method described in the last chapter, except that instead of just the total cost of the building being divided by the gross floor area to give a total cost per square metre, each one of the elements is also divided by the gross floor area to give a cost per square metre.

Figure 2 shows the cost analysis of a university student residence in which the costs have been analyzed this way. The gross floor area is 1696 m² and the total cost is $721 200, which gives a cost of $425.24 per m². Of this the STRUCTURE, for example, contributed $112 000 or $66.04 per m², and within the STRUCTURE the UPPER FLOOR CONSTRUCTION contributed $66 700 or $39.33 per m². Recorded against each element and sub-element is its total cost and its cost per square metre of the gross floor area. Note that the costs are all recorded to the nearest hundred dollars which is close enough for an analysis. It is also close enough for an estimate and helps guard against delusions of accuracy.

A budget estimate can now be made using this basic information, particularly if the known costs of several buildings have been analyzed in this way, and provided they are reasonably up-to-date. The gross floor area can be measured and, instead of having to select a cost per square metre for the whole building, a cost per square metre for each of the elements can be selected from the analyses and a total cost built up. It is always easier to build up a price than it is to select a cost per square metre for the complete building, and this also overcomes the principal disadvantage of the single-rate methods in that it is possible to allow in the estimate for differences in size, shape, quality and type of construction.

Another advantage claimed for elemental cost analysis is that it is possible to prepare the estimate without the benefit of drawings. In fact there was a move at one time advocating that estimates be prepared using elemental cost analyses before any sketch drawings had been developed, and because the estimate had been based on a known quality and type of construction, the drawings and specifications should then be prepared using the estimate as a guide. This means that design follows cost rather than the other way round. This approach might possibly work when applied to a building program such as a school building program where a number of buildings of a similar type have been analyzed and there is little variation in design, but it is not appropriate for most buildings. While there are many occasions when an estimate has to be prepared in advance of the drawings and specifications, following this up with the requirement that the design should be based on the estimate is rarely, if ever, suggested now, not only because it is not really very practical, but also, no doubt, because it is not popular with the design professions.

Although analyzing the cost of each element to give a cost per square metre for each is a step in the right direction, there are further analyses which can be made and which can be even more useful. One of the objectives in

ELEMENTAL COST ANALYSIS

Helyar & Associates
Helyar, Rae, Mauchan & Hall Limited
Chartered Quantity Surveyors - Construction Consultants

Project: University Students Residence
Gross Floor Area: 1696 m²
Architect: Travers Primi Associates
Project No. 78102
Date: March 1978
Estimate No. Analysis

ELEMENT	RATIO	QUANTITY	UNIT RATE	AMOUNT	COST PER m² GROSS	TOTAL
1. SUBSTRUCTURE						
a) Normal Foundations		628 m²	12.26	7700		
b) Basement		None	–	–		
c) Special Foundations		None	–	–		
Caissons etc. $						
Underpinning $						
Total					4.54	7700
2. STRUCTURE						
a) Lowest Floor Construction		628 m²	14.01	8800	5.19	
b) Upper Floor Construction		1068 m²	62.45	66700	39.33	
c) Roof Construction		628 m²	58.12	36500	21.52	
Total					66.04	112 000
3. EXTERIOR CLADDING	–					
a) Roof Finish		628 m²	20.38	12800	7.55	
b) Walls Below Ground Floor		None	–	–	–	
c) Walls Above Ground Floor		876 m²	56.51	49500	29.19	
d) Windows		261 m²	155.94	40700	24.00	
e) Exterior Doors and Screens		18 m²	194.44	3500	2.06	
f) Balconies and Projections		9 m²	88.89	800	0.47	
Total					63.27	107300
4. INTERIOR PARTITIONS AND DOORS						
a) Permanent Partitions and Doors		2363 m²	19.00	44900	26.41	
b) Movable Partitions and Doors		41 m²	80.49	3300	1.95	
c) Glazed Partitions and Doors		None	–	–	–	
Total					28.36	48200
				CARRIED FORWARD	$ 162.26	$ 275 200

FIGURE 3.

CONSTRUCTION ESTIMATING AND COSTING

Helyar & Associates

Page 2

ELEMENT	RATIO	QUANTITY	UNIT RATE	AMOUNT	COST PER m² GROSS	TOTAL
				BROUGHT FORWARD	162.26	275 200
5. VERTICAL MOVEMENT						
a) Stairs		144 m	60.42	8700	5.13	
b) Elevators and Escalators		None	—	—	—	
Total					5.13	8 700
6. INTERIOR FINISHES						
a) Floor Finishes		1408 m²	36.65	51 600	30.42	
b) Ceiling Finishes		1486 m²	15.07	22 400	13.21	
c) Wall Finishes (Wall Ratio ___)		4636 m²	17.75	82 300	48.53	
Total					92.16	156 300
7. FITTINGS AND EQUIPMENT						
a) Fittings and Fixtures				17 700	10.44	
b) Equipment				1 800	1.06	
Total					11.50	19 500
8. SERVICES						
a) Electrical				51 200	30.19	
b) Plumbing and Drains				33 200	19.58	
c) Heating, Ventilation and Air Conditioning				84 900	50.06	
Total					99.82	169 300
9. SITE DEVELOPMENT						
a) General		8465 m²	1.94	16 400	9.67	
b) Services				6 800	4.01	
c) Alterations				—	—	
d) Demolition				—	—	
Total					13.68	23 200
				CARRIED FORWARD	$384.55	$652 200

FIGURE 3. (continued)

ELEMENTAL COST ANALYSIS

Helyar & Associates

Page 3

ELEMENT	RATIO	QUANTITY	UNIT RATE	AMOUNT	COST PER m² GROSS	TOTAL
				BROUGHT FORWARD	384.55	652 200
10. OVERHEAD AND PROFIT						
a) Site Overhead				56 000	33.02	
b) Head Office Overhead and Profit				13 000	7.67	
Total					40.68	69 000
				SUB-TOTAL	425.24	721 200
11. CONTINGENCIES						
a) Design Contingency						
b) Escalation Contingency						
c) Post-Contract Contingency						
Total						
				TOTAL	$ 425.24	$ 721 200

COMMENTS:

FIGURE 3. (continued)

selecting the list of elements was that, wherever possible, each element should be capable of measurement, that is it should be possible to quantify the element. Figure 3 shows the next stage in analysis of the student residence in which quantities have been shown under the QUANTITY column. The way in which measurements are made for each of the elements is described in Appendix B.

When the amount of each element is divided by its quantity, the result is the unit rate shown in the UNIT RATE column. Although the amount is derived from a variety of materials and trades which make up the cost of an element, the unit rate gives an overall unit cost based on the quantity of the element, and also gives an overall indication of the standard of quality used. Care should be taken not to confuse the unit rate with the cost per square metre, the former being the amount divided by the quantity, while the latter is the amount divided by the gross floor area.

With a unit rate for most of the elements, if sketch drawings are available preliminary estimating becomes more accurate. If a set of block plans is available it is at least possible to measure the area of the foundations, the cube of the basement, if there is one, as well as the areas of the floors, roof and exterior cladding. Then instead of trying to estimate the cost of each element by using the cost per square metre of floor area, the elements themselves are measured as far as possible and their unit rates applied. For those elements which cannot be measured, the cost is derived from lump sum amounts, from the cost per square metre of the gross floor area or by use of the ratios.

Figure 4 shows the completed cost analysis sheet for the student residence with the ratios shown in the RATIO TO GROSS FLOOR AREA column. Wherever a quantity can be shown against an element it can be divided by the gross floor area to give a ratio, which can serve three useful purposes. First, it provides a means of checking for mathematical errors. For example, the ratios of LOWEST FLOOR CONSTRUCTION and UPPER FLOOR CONSTRUCTION should always add up to 1.0 because the quantities of these two elements when added together are equal to the total gross floor area. With experience of a number of analyses and estimates over a period of time the ratios of other elements become familiar, and if an unexpected ratio appears it would be wise to check the quantity to see if it is correct.

Second, the ratio will show why one building is more expensive than another. While the quality and type of construction of two buildings may be very similar, so that the unit rates for the elements are also very similar, the total costs may be quite different. A glance at the ratios will show why this is so; perhaps the more expensive building has a less efficient shape or contains proportionately more partitions, and the ratios will immediately indicate this.

Finally, the ratio can be used in very early estimates as a means of establishing a quantity. This applies particularly to partitions and finishes

ELEMENTAL COST ANALYSIS

Helyar & Associates
Helyar, Rae, Mauchan & Hall Limited
Chartered Quantity Surveyors · Construction Consultants

Project: University Residence
Gross Floor Area: 1696 m²
Architect: Tracey Prini Associates
Project No. 78102
Date: March 1978
Estimate No. Analysis

ELEMENT	RATIO	QUANTITY	UNIT RATE	AMOUNT	COST PER m² GROSS	TOTAL
1. SUBSTRUCTURE						
a) Normal Foundations	0.37	628 m²	12.26	7700	4.54	
b) Basement		None	—	—	—	
c) Special Foundations		None	—	—	—	
Caissons etc. $						
Underpinning $						
Total					4.54	7700
2. STRUCTURE						
a) Lowest Floor Construction	0.37	628 m²	14.01	8800	5.19	
b) Upper Floor Construction	0.63	1068 m²	62.45	66700	39.33	
c) Roof Construction	0.37	628 m²	58.12	36500	21.52	
Total	1.37	2324 m²			66.04	112 000
3. EXTERIOR CLADDING						
a) Roof Finish	—	628 m²	20.38	12800	7.55	
b) Walls Below Ground Floor		None	—	—	—	
c) Walls Above Ground Floor	0.52	876 m²	56.51	49500	29.19	
d) Windows	0.15	261 m²	155.94	40700	24.00	
e) Exterior Doors and Screens	0.01	18 m²	194.44	3500	2.06	
f) Balconies and Projections	0.01	9 m²	88.89	800	0.47	
Total (excluding (a) & (b))	0.68	1155 m²			63.27	107 300
4. INTERIOR PARTITIONS AND DOORS						
a) Permanent Partitions and Doors	1.39	2363 m²	19.00	44 900	26.47	
b) Movable Partitions and Doors	0.02	41 m²	80.49	3300	1.95	
c) Glazed Partitions and Doors		None	—	—	—	
Total	1.42	2404 m²			28.42	48 200
			CARRIED FORWARD		$ 162.26	$ 275 200

FIGURE 4.

CONSTRUCTION ESTIMATING AND COSTING

Helyar & Associates

Page 2

ELEMENT	RATIO	QUANTITY	UNIT RATE	AMOUNT	COST PER m' FORWARD	COST PER m' GROSS	TOTAL
				BROUGHT FORWARD		162.26	275 200
5 VERTICAL MOVEMENT							
a) Stairs	0.08	144 m	60.42	8 700	5.13		
b) Elevators and Escalators		None	—	—	—		
Total						5.13	8 700
6 INTERIOR FINISHES							
a) Floor Finishes	0.83	1408 m²	36.65	51 600	30.42		
b) Ceiling Finishes	0.88	1486 m²	15.07	22 400	13.21		
c) Wall Finishes (Wall Ratio 0.79)	2.73	4636 m²	17.75	82 300	48.53		
Total						92.16	156 300
7. FITTINGS AND EQUIPMENT							
a) Fittings and Fixtures				17 700	10.44		
b) Equipment				1 800	1.06		
Total						11.50	19 500
8. SERVICES							
a) Electrical				51 200	30.19		
b) Plumbing and Drains				33 200	19.58		
c) Heating, Ventilation and Air Conditioning				84 900	50.06		
Total						99.82	169 300
9. SITE DEVELOPMENT							
a) General		8465 m²	1.94	16 400	9.67		
b) Services				6 800	4.01		
c) Alterations				—	—		
d) Demolition				—	—		
Total						13.68	23 200
				CARRIED FORWARD		$384.55	$652 200

FIGURE 4 (continued)

Helyar & Associates

Page 3

ELEMENT	RATIO	QUANTITY	UNIT RATE	AMOUNT	COST PER m²	GROSS	TOTAL
				BROUGHT	FORWARD	384.55	652 200
10. OVERHEAD AND PROFIT							
a) Site Overhead				56 000	33.02		
b) Head Office Overhead and Profit				13 000	7.67		
Total				SUB-TOTAL		40.68	69 000
						425.24	721 200
11. CONTINGENCIES							
a) Design Contingency							
b) Escalation Contingency							
c) Post-Contract Contingency							
Total							
				TOTAL		$425.24	$ 721 200

COMMENTS:

FIGURE 4. (continued)

when a preliminary drawing may indicate the overall size and shape of the building, but doesn't show how the interior will be sub-divided. The ratios contained in an analysis of a similar type of building can be used to provide the missing quantities which, while they won't be exact, will be reasonably close. It might also be noted that the ratio multiplied by the unit rate equals the cost per square metre gross. This means there is an alternative way of calculating the amount. Instead of finding the quantity by using the ratio and then multiplying it by the unit rate, multiply the ratio by the unit rate and then multiply the result by the gross floor area to arrive at a total cost for the element.

The Uses of an Elemental Cost Analysis

An elemental cost analysis has several uses, most of which are not found in other methods of cost analysis. It shows how the costs are distributed over the building, and in doing so it shows up the importance, as far as the cost is concerned, of any one element in the total. A review of the figures in the COST PER m² GROSS column indicates immediately whether an element represents a major part of the cost of the building or whether, in fact, its cost is comparatively insignificant.

It also enables a comparison to be made between the costs of the same element in two or more similar buildings, and to see whether money might have been unnecessarily spent on any of the elements. This is done by comparing the cost per square metre of the gross floor area for an element in each of the buildings. Because the costs of the elements have been reduced to costs per square metre buildings of different sizes can be compared provided they are of the same type. The interior finishes of the student residence, for example, amount to $92.16 per m² and comparison with similar types of buildings shows that this is high. This might have been pointed out to the architect during design and he would then have had the opportunity to make the necessary corrections to bring the costs back into balance, or to provide a valid reason for the apparent imbalance. In any case, whenever possible attempts should be made to balance the costs so that the building owner obtains the best value for his money. However, a number of analyses of similar building types are required before a reasonable balance of costs between the elements can finally be established.

Lastly, an elemental cost analysis provides useful information for estimating and for planning the cost of future projects, although it needs to be employed with a certain amount of caution. Every building is unique, site conditions and market conditions are variable, and recorded costs soon become outdated, which means that while the cost analysis can be useful as a guide, a certain amount of common sense together with a knowledge of current construction costs are needed in its application for forecasting the cost of a proposed building.

CHAPTER 5

Cost Planning

Cost control in the pre-tender period of a construction program has three main objectives: to ensure that the actual cost of the building does not exceed the owner's budget, to give the owner good value for his money and to achieve a proper balance of expenditure over the various building components. If the first of these objectives can be achieved the others will often follow, particularly if the elemental cost method is used.

PROGRAM STAGE

In order to see how the elemental method can be used to plan and control the cost it is useful to do so in the context of the design process in the pre-tender period. The design process can be seen as a series of stages, often overlapping, which lead up to the final set of working drawings and specifications from which the contractors can estimate their costs in order to submit tenders. The first of these stages is the program or concept stage. At this stage the building owner, depending on what type of owner he is, will have made his first feasibility study, will appoint consultants to start an investigation into his requirements or will approach an architect with an outline of what he wants based on his own investigation. Whatever form it might take the result is some description, perhaps a written program, which will indicate the type of building to be built, its location, possibly its overall size, its standard of quality and the types and net areas of spaces which have to be provided within the building. If any drawings are prepared they would only show block plans, probably at a small scale, with perhaps some indication of how the interior might be planned in very broad outline.

It is quite possible that the building owner would want to discuss costs at this stage. He may have already determined how much he can afford to spend and wants to know whether the building he has in mind can be built within his budget, or he has not yet established his budget and wants to know how much money should be allocated to his project. In either case an estimate of costs has to be made.

This situation raises an interesting paradox. The first estimate given to a client is possibly the most important estimate in the life of the project. It is the estimate the client will always remember because it provides the cost information which will be the basis of a major financial decision, including the decision as to whether it is worth going ahead with the project. It is also important to the design consultants because an inadequate estimate at this stage will cause them a great deal of trouble in the future. The first estimate should, therefore, be as accurate as possible: not so low as to mislead the building owner into proceeding with the project, nor so high as to cause it to be cancelled when it was in fact perfectly viable. At the same time it is the estimate which is based on the least amount of information. The building owner may have a reasonably clear idea of what he wants in his building in terms of accommodation, function and standard of quality, but these are still a long way from being translated into construction terms of bricks and mortar, which is what costs are based on.

Although the first estimate must be as accurate as possible, if it is asked for in the program stage it cannot be prepared in great detail because of the lack of information available at that point. A program-stage estimate has to be based more on assumptions than on facts, unlike later estimates which can be based on facts with only a few assumptions. It will tell the client what it is possible for the building to cost, but not necessarily what it will actually cost, and he can hardly expect a definitive statement of cost at this time. Most clients recognize this and, provided they are kept informed of their revised cost commitments during the subsequent design stages, will accept minor variations from the first estimate. It is also worth noting that although construction costs are very important to any client, a small rise in mortgage interest rates can upset a client's budget to a much greater extent than a small increase in construction costs. Very often, too, an upgrading of quality in the later stages of the design of the building, although it will increase construction costs, may give the client a more acceptable building than he had first visualized or enable him to charge higher rents to offset the increased cost. The first estimate should therefore be prepared in a responsible manner but not treated as a sacrosanct figure which is unchangeable.

Since drawings are not usually available at the concept stage the estimate must be prepared from the program. If the gross floor area is not given it has to be calculated from the net areas given in the program, using net to gross ratios obtained from previous buildings of a similar nature, and the total cost calculated either by the square metre method, or by building up the cost using costs per square metre for each of the elements obtained from available elemental cost analyses as was described in the last chapter.

Whichever method is used, the following adjustments may have to be made to the estimate:

1. TIME

Analyses of previous buildings soon became obsolete and unless the analyses which are being used for the estimate are very recent some means will have to be found to update them. A cost index is the usual means of doing this, but most indices which are intended for construction are input rather than output indices. Statistics Canada, for example, publishes a labour cost index as well as a material cost index, and by combining the two, a total cost index for both residential and non-residential construction, but these figures are Canada-wide and are based on the costs which contractors have to pay for labour and material. This might be of help to a contractor by telling him how much his costs are rising, but since they don't reflect changes in productivity or market conditions at the general contractor level they are not of much assistance in updating an estimate. What is really needed is an index showing changes in costs to the building owner, an output index, rather than changes in costs to the contractor. Besides Statistics Canada, other published indices are the *Southam Construction Cost Index* and the index in the *Boeckh Building Valuation Manual* which cover costs in Canada, and the index in *Engineering News Record* which covers costs in the United States. The last two both purport to be output indices.

2. SIZE AND SHAPE

The effect of size, shape and the number of storeys on the cost have already been mentioned, as has the fact that it is easier to make adjustments when the cost is built up by elements. However, the lack of drawings at the program stage makes the adjustment for these three factors difficult even with an elemental breakdown, and although the adjustment has to be made it becomes a matter of guesswork allied to experience.

3. QUALITY

Again adjustments for changes in quality are easier to handle with an elemental breakdown. Also it is easier to explain to the client how, and to what extent, these adjustments have been made than it is with an estimate prepared by the square metre or any other single-rate method.

4. LOCATION

Costs will vary from location to location and if the proposed building is in one city while the only available cost analyses are for similar buildings in a different city an adjustment needs to be made. Publications which give index numbers between cities in North America include Means *Building Construction Cost Data* and the *Boeckh Building Valuation Manual*.

5. OTHER FACTORS

These include differences in soil and site conditions, inclusions and exclusions and contingencies. Such matters as whether the cost analyses include or exclude site works, full partitioning, all fittings, fixtures, equipment, etc. must be considered. The cost analyses must be studied carefully to ensure that they are complete. It is essential on the one hand that they cover all the costs which should be included in the estimate for the proposed building, while on the other hand care must be taken not to include more than is necessary. Costs such as fees and other development expenses might also come under this heading, but it should be noted that they are not normally included in the estimate, the estimate being intended to cover normal construction costs only.

Even though all the foregoing adjustments have been made it is wise to add a contingency at the end of the estimate to cover unforeseen problems. An estimate prepared at this stage, although it may have been done with great care, cannot possibly cover all the revisions which will take place between concept and final contract documents. It is therefore essential to include a contingency. Depending upon the presumed accuracy of the estimate and the adjustments which have been made this could vary between five and ten per cent. The subject of contingencies is dealt with in more detail in Appendix B.

At the completion of the program stage the first estimate can be presented for the client's approval and acceptance. It should be an honest attempt to show what the building might cost, based on the available facts, and should not be influenced by the client's (or anyone else's) uninformed opinions about costs. Infinite trouble can develop later if the estimate is pared because it is believed that a lower estimate will be more acceptable to the client, and it is patently dishonest to deliberately distort the estimate with the object of securing a commission. Accompanying the estimate should be a brief description of what it does and does not include. This not only provides the client with needed information but also serves as useful documentation in the later stages of design.

SCHEMATICS STAGE

Following the program stage is the schematics stage when the design team starts addressing itself more toward the construction with the production of preliminary design drawings and an outline specification of the building. If an estimate of cost was not asked for during the program stage it will certainly be requested at this point, but because more information is now becoming available the accuracy of the estimate can be more certain. During this stage the architect will produce sketch drawings which show, at the very least, the basic size and shape of the building from which exterior wall, floor and roof areas can be measured, together with a specification which outlines the general standard of quality. The structural engineer will give some thought to the structural system he intends to use, and the mechanical and electrical engineers will do the same for their systems. The information will not be very detailed, but at least it is more than was available during the program stage.

This information does not become available in one neatly prepared package, but usually in a constantly flowing stream throughout the duration of the schematics stage. Floors can be added here and taken away there, the shape changed from square to rectangular and back again and the interior planning amended with revisions being made to the specification almost on a daily

COST PLANNING

basis. Out of this welter of activity should come one or more schemes which need to be costed and compared with the estimate, if there was one, prepared during the program stage.

The estimate during the schematics stage must be prepared in close collaboration with the architect, his engineers, any other consultants and, whenever possible, with the client, and all must take some responsibility for its accuracy. Depending on the amount of information available, the following is an outline of the approach which might be adopted:

1. SUBSTRUCTURE

It is highly unlikely that the structural engineer will have designed the foundations at this stage. The only way of estimating NORMAL FOUNDATIONS will be by measuring the "footprint" of the building and using cost analyses of buildings which have the same number of storeys and similar soil conditions. BASEMENT can be measured either as a cube as described in Appendix B, or the actual quantities of excavation and backfill can be measured and priced separately.

2. STRUCTURE

Measurements can be made as described in Appendix B, but the structural engineer should be consulted about the type of construction and how it will vary in different parts of the building. He is usually very helpful in giving information on the probable mass of structural steel or reinforcing steel per square metre of the structure, the column spacings, and the possible thickness and types of floor and roof slabs. Usually, rather than relying on cost analyses for the information, it is better to work out the cost of typical bays from information provided by the structural engineer, as was described in Chapter 3.

3. EXTERIOR CLADDING

This can usually be measured, although it may be necessary to work out the areas of WINDOWS and EXTERIOR DOORS AND SCREENS as a percentage of the WALLS ABOVE GROUND FLOOR. The architect may have some idea of what he proposes to use for the exterior cladding, or he may only have a general idea of the standard of quality. In the former case it is possible to build up unit rates for each of the sub-elements, but in the latter case reference would have to be made to cost analyses and unit rates obtained from them. In either case the intent is not to pin the architect down irrevocably to any particular material or system, but only to ensure that there is a reasonable amount of money in the budget to allow him some options in the later design stages.

4. INTERIOR PARTITIONS AND DOORS

If partitions and doors are shown on the drawings they can be measured and priced. However, usually they aren't shown, or only some of them are shown, or the partitions are shown without the doors, and in any case nobody knows quite what type of partitions will be used. Probably the best solution is to use the ratios from cost analyses to give quantities for the partitions and doors, or if they can be measured, to use the ratios to check the quantities. Remember that if only some of the partitions are shown there is no point in measuring them because it still leaves the remainder to be guessed at, and if an unknown remainder has to be guessed at it is better in the long run to calculate all of them by means of the ratio. Pricing of INTERIOR PARTITIONS AND DOORS, because they are always such a mixture of different types, is best done by using the unit rates in cost analyses of similar building types.

5. VERTICAL MOVEMENT

STAIRS can be calculated by estimating the total length of risers and applying a unit rate, or by allowing lump sum costs for each flight. ELEVATORS AND ESCALATORS can also be costed on a lump sum basis.

6. INTERIOR FINISHES

Quantities can be established by ratios and quality by percentages of types of finish. Interior finishes are always a time-consuming measurement and one that is out of all proportion to their value in the total building cost, particularly in the case of WALL FINISHES. On the theory that wall finishes are applied to the inside face of the exterior walls and to both faces of the partitions, a quick way of measuring wall finishes is to take once times the quantity of WALLS BELOW GROUND FLOOR, plus, once times WALLS ABOVE GROUND FLOOR, plus twice times PERMANENT PARTITIONS AND DOORS, and then, because duct shafts and ceiling spaces are unfinished, apply a factor, say 75%, to the result to give an approximate quantity of wall finishes. The factor of 75% will vary depending on the type of building, but if this is recorded with each cost analysis it can be used for estimates at the schemastics stage.

7. FITTINGS AND EQUIPMENT

Quantities cannot be measured and it is unlikely that the architect has given much thought to this element at such an early stage. The estimate must therefore be based on costs per square metre of the gross floor area obtained from cost analyses, with adjustments for any known items of fittings, fixtures or equipment.

8. SERVICES

Electrical and mechanical engineers seem to follow along even further behind the architect than structural engineers, and normally it would not be possible to measure quantities for SERVICES. The costs therefore have to be estimated on the basis of square metre costs, either as a total for each sub-element, or further broken down into systems as described in Appendix

COST PLANNING

B. It is advisable to review the estimate with the engineers before becoming committed to it.

9. SITE DEVELOPMENT
If any of the site development is shown it can be measured and priced as described in Appendix B, otherwise it should be priced as an overall cost per square metre of the site since the cost usually bears no relation to the area of the building. GENERAL SITE DEVELOPMENT is something of a rubber-band item since the cost can be expanded and contracted, within limits, without affecting the quality of the building. ALTERATIONS and DEMOLITIONS, if they occur, have to be treated separately from other site development costs and included as lump sum estimates at this stage.

10. OVERHEAD AND PROFIT
This can be included as a percentage of the construction cost, the percentage having been taken from previous cost analyses.

11. CONTINGENCIES
These are described in Appendix B. It is essential to have a design contingency at this stage so as to give the design consultants some flexibility in their design decisions during the later stages. ESCALATION CONTINGENCY may or may not be included, some clients preferring to use their own crystal ball, but if it is excluded it should be so stated when the estimate is presented. Similarly, some clients would rather include the POST-CONTRACT CONTINGENCY as part of their overall budget and not as part of the construction estimate.

It is very unusual for only one estimate to have to be prepared during this stage. Normally three or four are required so that different solutions can be compared, or because the architect believes he can improve on the previous schemes. However, by the end of the schematics stage all the fundamental decisions, certainly those concerning cost, have been made. The size, shape, height, type of construction and general standard of quality will have been established and, although detailed design decisions may not yet have been made, these key decisions will have determined the basic cost of the building. Unless a major change is made in the overall concept, all future decisions will have comparatively minor effect on the cost. It is important to remember that the irreversible decisions regarding cost are made by the end of the schematics stage, not at some later phase in the design.

The finally accepted estimate in the schematics stage now becomes more than an estimate. It is an interpretation in costs of all the key decisions which have been made by the design team, including the client, in a form which shows how it is intended to spend the money on each one of the elements in the building. It is in fact a cost plan which, if followed correctly, will ensure that the budget is not exceeded.

The reason for introducing the term "cost plan" rather than "estimate" is that there is a subtle difference between the two. An estimate is usually thought of as a prediction of cost based on a particular design, and the estimate can only be changed if the design is changed. A contractor's estimate, for example, contributes nothing to the design, it only reflects what a particular design will cost after all the design decisions have been made, and its function is therefore passive. The process of preparing estimates during the schematics stage on the other hand contributes to the design by showing the cost effect of alternative decisions, and by allocating costs to elements (without nailing down precisely the materials and methods making up those costs) it does not commit the design team to irrevocable future decisions. It demonstrates how costs might be expended but does not specify in detail how they must be expended, thus allowing for flexibility within the overall cost. At the same time as the estimates culminating in the cost plan are being prepared questions should be raised about whether components are making the best contribution to the building, whether there is a better way of incorporating them or whether they need to be there at all. The function of the cost plan, both in its preparation and in its use later in the design process, is therefore active.

Because it is going to be used as a working tool, the presentation of the cost plan is important. It should show the cost breakdown by elements and be accompanied by an outline specification describing what has been included. In many instances it is useful to supplement this with a complete cost breakdown showing the quantities and unit prices which have been measured or assumed to establish the cost plan.

Each member of the design team should be given a copy of the cost plan together with all necessary supporting material so that they know the parameters within which they will be working as the design is developed and how those parameters were established. Since the design team played a major role in setting the parameters there should be no difficulty in accepting them, and the conclusion of the schematics stage should be a basic scheme coupled with a cost plan, both of which are acceptable to the client and his design consultants.

CHAPTER 6
Cost Control

The theory of cost planning described in the last chapter can be applied without much difficulty. The theory of cost control described in this chapter is not so readily put into practice, possibly because it requires the design consultants to follow a more formalized procedure than they are used to. Cost planning really imposes no new requirements on the design team up to the end of the schematics stage, but cost control introduces the concept of designing by elements and some architects may have difficulty in accepting this. Also, the distinction between design development and contract documents as it is described here can sometimes become blurred. When these concepts are not followed it is usually because cost control is not fully understood or accepted by the design team with the result that decisions are sometimes made without reference to the cost plan, in which case costs are not properly controlled and the time spent in preparing the cost plan has been wasted.

The essence of any cost control program is to have a frame of reference, followed by a means of checking, and finally a means of taking remedial action if costs are getting out of control. In this instance the cost plan is the frame of reference. It shows the total amount of money to be spent, it shows how the money is to be distributed over the elements and thus sets up a target cost for each of them. It should also show a properly balanced distribution between the elements. A cost plan is essentially similar to any other budgeting program, as, for example, a household budget in which there is a total amount of money allocated, a reasonable distribution, but no fixed rules dictating exactly how it will be spent.

DESIGN DEVELOPMENT STAGE

During the design development stage the design team will consider the detailed design of all the components of the building. To illustrate how the cost plan is used to help control cost consider the element WALLS ABOVE

GROUND FLOOR. At the schematics stage it is possible that no firm decision was made on the type of exterior wall which was to be used, only a general assumption perhaps that some form of precast concrete cladding of a certain quality might be considered. The cost plan would reflect this, showing that an allowance of so much per square metre for a cladding system of this quality was included. Now, at the design development stage, a firm decision does have to be made on the cladding, but before making his final decision the designer might consider three or four alternatives. In doing this he might prepare the necessary preliminary design drawings and other information for each of the alternatives and have them costed using the unit in place method. The costs would then be tabulated to assist the designer in making the final selection. This is the means of checking described earlier.

Of the alternatives under consideration, some may be more acceptable architecturally but be too expensive; others may meet the budget but be less acceptable as a design solution. The ideal would be a design which is both architecturally desirable and within the cost target for the element. However, if it appears that insufficient money was allocated in the cost plan to give the designer what he wants, and this does sometimes happen, three courses of action are available:

1. He can accept the higher figure and make a compensating reduction in one or more of the other elements to cover the cost difference. He may already have found the cost plan too high for another element, in which case the money is available from that element. If he hasn't already found savings in the cost plan the first element to examine as a possible source of additional money is the design contingency, although there may be other elements the designer feels are likely to contribute something toward the overexpenditure. The effect of this procedure, if it can reasonably be accomplished, is to modify the distribution of costs shown on the cost plan, while still maintaining the overall budget.

2. He can resign himself to accepting a less desirable architectural solution to keep within budget. Although this may not be entirely satisfactory to the designer, the original cost plan is maintained. This course of action is the only one available if the other two are impractical or unacceptable.

3. He can ask his client for additional money to cover the extra cost. This is the least desirable course of action and because it is unlikely to be a satisfactory solution from the client's standpoint should only be considered in very exceptional circumstances. However, at least the client is warned that the budget may need to be increased reasonably early on in the design process. An explanation of why more money is needed should also be given to help the client make his decision. If the final conclusion is to increase the budget this will also affect the distribution of costs over the elements.

Accepting one of these three options is the remedial action mentioned earlier.

All the elements are considered within this context until an acceptable solution is found for each of them in a sequence of:
1. Design of the element
2. Costing of the element
3. Comparison with the cost plan
4. Decision and action on the design.

It is often useful to have a standard form, not unlike a change order form, to record each of the cost checks as they are approved, showing the revised figure for the element; the effect, if any, on other elements; and the revised amount of design contingency to keep the total cost on budget. An example of such a form is included in Appendix D.

At the end of the design development stage, designs of all the elements will have been completed and the cost of each element tested against its target cost. There may have been a slight redistribution of the costs over the elements, but provided the cost plan was adequate change, if any, in the overall cost will be minor.

CONTRACT DOCUMENT STAGE

The final phase of the design process is the contract document stage when the decisions made at the design development stage are translated into working drawings and specifications. As there is a tendency for these two stages to overlap it is advisable to prepare a complete estimate using the unit in place method early on in the contract document stage. This helps to ensure that a decision taken on one element has not affected the cost of another element, a situation which may arise and pass unnoticed when elements are being costed one at a time. In practice the cost checking of elements in the design development stage does not proceed as neatly as the process has been described here so it is unlikely that costs will be overlooked but it is still usually well worth while preparing a full estimate early in this stage.

A further estimate, also using the unit in place method, when the working drawings are about 80% complete is also recommended to make sure that no additional cost items have manifested themselves while the working drawings and specifications are being developed. This estimate is customarily done on an element breakdown and is really a final updating of the cost plan. It can be used as a cost analysis for help with future projects but should also be re-analyzed in terms of a trade breakdown so a comparison can be made between the estimate and the breakdown submitted by the successful contractor. This serves two purposes: it helps ensure that the contractor's breakdown is reasonably accurate, and it can show where discrepancies exist between the two estimates so that the elemental cost analysis can be adjusted, if necessary, to make it reflect the actual tender more accurately.

In the case of a management contract, or fast-track, the same basic sequence of events takes place, the principal difference being that the trade breakdown is prepared earlier on in the process. An elemental cost plan is still required at the end of the schematics stage as an aid to the design team, although the trade breakdown is started shortly thereafter as trade packages are identified for tenders to be compared with the budget. The trade breakdown then becomes the cost plan for construction. As tenders are called and checked against the construction cost plan, adjustments may have to be made to the design and the specifications to bring expenditure back within the budget. Although a management contractor is brought in at an early stage some management contractors have experience only of costs in the general contractors' trades and for this reason it is usually preferable to have the individual who prepared the original preliminary estimate and cost plan continue with the cost control until all tenders have been called and the final tender cost known. It also helps to ensure continued responsibility.

Figure 5 shows the cost planning and cost control process in chart form. It should be made clear that this process is not intended to produce cheap buildings, nor is it intended to inhibit design. The aim is to ensure that from the outset the building owner's budget is adequate for the type of building contemplated, that he gets the best value for his money, and that he doesn't receive an unpleasant shock when the tenders are opened. The designer, in the meantime, should be kept informed of the cost effect of all his decisions and be given adequate warning that costs are running high before he becomes too involved in his design to want to give it up. Too often the inadequacy of the budget becomes evident only when it is next to impossible to do anything about it. As for inhibiting the design, there is nothing more inhibiting to a design than an inadequate budget and a lack of knowledge of the cost implications of design decisions.

As was mentioned at the beginning of this chapter, perhaps the principal reason for the discrepancy between cost control in theory and in practice is that many architects and their consultants don't really design by elements. For a cost control program to really work all members of the design team — the project manager, the architect, the engineers, the other consultants and the building owner must all be fully committed to it and take full responsibility for making the program work. The architect and his engineers should understand the cost control process and make any adjustments in their methods of production that may be necessary. Only then, with the architect and his consultants organizing their information to co-ordinate with the cost plan, with everyone accepting the reliability of the costs and recognizing that cost control is the responsibility of the whole team, will cost control begin to work.

At the same time, it is important that the subject be kept in perspective. Estimating, cost planning and cost control are all exercises in forecasting, and forecasting the future, as any good weatherman knows, is an acceptable

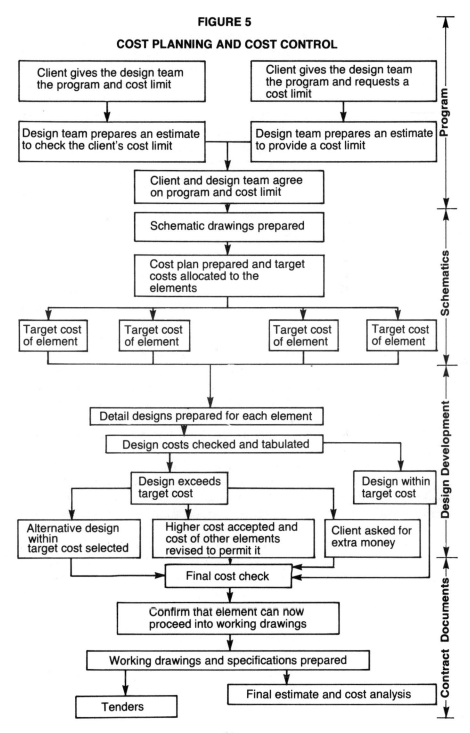

FIGURE 5

COST PLANNING AND COST CONTROL

pastime provided it isn't done more than six hours ahead. A question which is frequently raised is how close tenders should be to the final estimate or cost plan. With a spread of ten per cent or more between bidders on many projects it is to be hoped that the difference between the cost plan and the low tender should not be more than five per cent, particularly if design and cost can be co-ordinated in the manner described. However, although the design team can call on a substantial body of past experience and can assess current trends there is no way the results can be guaranteed. Usually the results are satisfactory but sometimes through unforeseen circumstances a disaster can occur. The only consolation available then is that it might have been considerably worse if cost planning and cost control had not been employed.

PART 2

Economics of Development

CHAPTER 7
Feasibility Studies

Until quite recently many real estate developments have been carried out by developers with little more than a feeling for a particular site and the drive and endurance to push the project through to completion with a minimum of their own money and practically no professional advice. Frequently the economic viability of the development was justified with a page or two of calculations based on questionable facts provided by people who had a vested interest in seeing the project started. And more often than not the developers had no idea of what their real costs were until some time after the development was completed. Inevitably any correlation between the final cost and the developers' original calculations was purely coincidental. Many of these developers have since disappeared from the scene and the fact that some of them are still in business is because inflation took care of their mistakes.

In recent years the emergence of a more sophisticated developer and the large size of some of the urban developments now being undertaken, not to mention the anti-development citizen groups and municipal politicians, has meant that the days of the developer who does not make a thorough preliminary study are numbered. A fully comprehensive feasibility study can require the combined talents of a tremendous number of people including architects, engineers, space planners, market analysts, appraisers, realtors, traffic consultants, lawyers, accountants, money managers, builders, quantity surveyors, sociologists, economists, environmentalists and ecologists.

Stated in its simplest terms there are two criteria by which the feasibility or economic merit of a development will be judged. These are the two factors mentioned in Chapter One, cost and value, and it is the relationship between them which is analyzed in a feasibility study.

While a feasibility study attempts to show that the value of a real estate development at least equals its cost, not all building owners concern themselves with the economic facts alone. An institutional owner may be more concerned with the design of the building as a prestige symbol, conferring indirect benefits far beyond what analysis shows is its pure economic worth, in which case the concern is not for a building which has the highest value in terms of being a good money-maker. Instead the building being sought is one which has the highest value as a prestige symbol, while at the same time being the least money loser.

Although feasibility studies are usually thought of in the context of new construction they can also be applied to alterations and additions to existing buildings or to determining whether old buildings should be renovated or demolished. When an existing building is to be enlarged a feasibility study should be made on the before and after values of the property to see whether the cost of the addition is viable. Similarly, a feasibility study can assist a building owner in determining whether it would be more economically sound to renovate an existing building than to pull it down and rebuild. In this instance the outcome could possibly be that the income in either case would be less than could be earned by selling the site and investing the proceeds of the sale in some other form of investment.

In most circumstances a feasibility study is directed toward providing investors with a measure of confidence in the probable economic results of investing their money in a real estate development. In other words, whether they will receive a good return on their investment. The type of real estate of interest to such investors would be self-sustaining, revenue-producing properties such as office buildings, apartment buildings, shopping centres, medical centres, industrial buildings, and, in some instances, housing developments.

MARKET STUDIES

The first step in the preparation of a feasibility study is a market survey to determine whether there is a demand for the type of property the developer intends to build. This is quite a critical stage since an improperly conducted market study can make nonsense of all that follows. If there is an obvious demand the study may be done by the developer, but more often than not it is advisable to enlist the services of a consulting firm specializing in this type of work. No matter who is responsible for the study it is important to remember that merely knowing accommodation shortages exist does not in itself constitute a demand; it has to be accompanied by a willingness on the part of potential tenants to pay rents at levels which make the development economically viable. Demand signifies that which can be paid for.

The market study would cover such matters as:
1. The amount of comparable accommodation presently available, the extent to which it is being rented, and whether the vacancy rate is high.

2. Whether there might be competition for the development from any other new projects being contemplated.

FEASIBILITY STUDIES

3. The rate at which demand for new space has been satisfied in recent years, and the probable demand in the future based on possible business expansion and economic growth.

4. The prevailing rents in the area, together with an assessment of local leasing practices with regard to escalation, renewals, length of term, etc.

5. An assessment of transportation facilities available in the area.

The market study will provide essential information upon which the developer can make initial decisions. The research may indicate that the development could be enlarged or reduced in size, developed in stages, revised in concept, postponed or abandoned altogether; decisions which it is wise to make as early as possible in the development process.

CAPITAL COSTS

If the market study indicates that the development might be viable the next step is to assemble the capital costs which will be required to complete the project. Capital costs are often referred to as either hard costs or soft costs. Soft costs are also sometimes called development costs. Hard costs would include the following items:

1. LAND

This would include not only the actual cost of the land but also real estate commissions, title insurance and land transfer taxes. Where the developer has owned the land for some time it should be appraised at its current value and this amount used. If the land is to be leased, its capitalized cost is used.

2. DEMOLITION

If existing buildings have to be demolished, or even if only a tree is to be removed, these costs have to be included.

3. CONSTRUCTION

This is the largest single cost and it is therefore crucial. Many developers include in this category only the actual cost of construction, while others include architect's and other consultant's fees.

Soft costs, or development costs, would include the cost of:

1. LAND SURVEYS

Two surveys must be conducted, one prior to construction and another when the development is finished. These are required for both construction and mortgage purposes.

2. SOIL TESTS

3. MARKET STUDIES
These have been described in the previous section of this chapter.

4. APPRAISAL FEES
This would include the fees for the mortgage company's appraisal as well as any appraisals done by the developer.

5. DESIGN FEES

6. SPECIAL CONSULTANTS' FEES
These cover specialized professional consultants such as traffic consultants, acoustic consultants and any other consultants whose fees are not included elsewhere.

7. LEGAL FEES
These would include fees in connection with the purchase of the land, incorporating a property company to handle the development, mortgage company legal costs, developer's mortgage legal costs, land transfer charges, re-zoning applications, drafting leases, trust deeds, and bond purchase agreements, etc.

8. ACCOUNTING FEES

9. UNDERWRITERS' FEES
If the money for the development is to be raised by means of a bond issue the underwriters will require a fee.

10. DEVELOPER'S FEE
The developer may require a fee over and above any return he makes on his investment in the property.

11. PROJECT MANAGEMENT FEE

12. FURNISHINGS
This may be quite minimal on some developments. It would include loose furniture, sun drapes, carpeting and any other items needed to make the building habitable and which have not been included in the construction cost.

13. ART WORK
If this has not been allowed for in the construction cost it should be included as a soft cost.

FEASIBILITY STUDIES

14. MUNICIPAL LEVIES
These are capital levies imposed by some municipalities on new developments as a contribution toward municipal services such as sewers, etc.

15. PARKING LEVIES
If the development does not provide sufficient parking the municipality may levy a charge to help defray the cost of public parking.

16. CAPITAL TAXES
Some provincial governments levy a tax on paid up capital stock, retained earnings, loans, etc. of incorporated companies.

17. REAL ESTATE TAXES
These are levied on the land before construction and on the development during construction. An improvement tax during construction may be levied on a phased development where part of the building is handed over and occupied prior to general completion.

18. INSURANCES
On new construction the general contractor is usually required to take out an all-risks insurance policy on the construction. However, if for some reason, this is not being done the cost of the insurance premiums during construction should be included as a soft cost.

19. MORTGAGE INSURANCE
This protects the mortgage company against the developer defaulting in the mortgage payments.

20. PERFORMANCE BONDS
These should be included as a soft cost if they have not been accounted for in the construction cost.

21. INITIAL MAINTENANCE COSTS
Although the contractor may give a twelve-month warranty on his work and materials there is likely to be some maintenance work required between the completion of construction and the start of income from the development.

22. ACCOMMODATION
Space may have to be rented to accommodate the developer's staff and consultants such as the project manager.

23. DEVELOPER'S ADMINISTRATIVE COST
These may be included in the Developer's Fee (10) but if they aren't, such

items as salaries, heat, light, telephone, travel, and other administrative expenses should be entered here.

24. LEASING COMMISSIONS
The developer may elect to arrange his own leasing but if a real estate firm is brought in as the leasing agent fees will have to be paid for this service, usually based on a percentage of the rental income.

25. ADVERTISING AND PUBLIC RELATIONS

26. TENANT INDUCEMENTS
As an enticement to potential tenants the developer will frequently offer an allowance of a certain amount per square metre toward the cost of tenant partitions and finishes.

27. TENANT LEASE TAKEOVERS
As a further inducement to a potential tenant the developer may agree to take over an existing lease if he moves into the new building.

28. INTERIM LENDER COMMITMENT FEES
In periods of tight money a small percentage may be levied by the interim lender on that portion of the loan which has not been drawn down by the developer.

29. LONG-TERM LENDER STANDBY FEES
This is similar to the previous item but would be levied by the mortgage company on the mortgage loan.

30. BROKERAGE COMMISSION
This is a finder's fee paid by the developer for being introduced to sources of money.

31. REGISTRATION FEES
If money is to be raised by means of debentures the charge for registering them has to be included.

32. INTERIM FINANCE
This is sometimes known as bridging financing and takes the form of a short term loan to cover expenses until the permanent financing can be arranged. Since permanent financing is not usually provided until construction is complete, and since interim financing has a much higher rate of interest, the amount which is included here depends to a large extent on the efficiency and speed of the design and construction teams. A long period of construction can add enormously to the cost of interim financing.

FEASIBILITY STUDIES

33. INITIAL OCCUPANCY COSTS

These are initial costs incurred between the completion of construction and the occupancy of the building such as the cost of training staff to operate the building.

34. CONTINGENCIES

This is an allowance to cover all the unforeseen costs arising from items which have been underestimated or overlooked in estimating the total capital cost.

35. INCOME DURING CONSTRUCTION

The feasibility study has to be based on figures current at a certain point in time, usually when construction is finished. If there is any likelihood that some income will be generated before construction is completed as, for example, in a phased development, this income should be capitalized and deducted from the total capital cost.

Not all these costs will be applicable to every development but they nevertheless function as a useful checklist for most of the expenses which can be expected on any major development.

INCOME

Since a feasibility study is directed toward showing that the real estate value of a development is at least equal to its costs the next step is to examine the value, for which the potential income and running expenses are required.

The sole source of income from an income-producing real estate development is the rentals which can be charged, and one of the principal tasks of the design consultants is to maximize this revenue. However, this objective should not override all other considerations, the attempt rather should be to achieve a proper balance between income and operating expenses, with due regard being paid to those features which contribute to the preservation of the investment on a long-term basis, not to mention those which contribute to the environment.

Where the feasibility study is being done for the purpose of deciding the worth of purchasing an existing revenue-producing property, the assessment of income is based on the existing revenue generated by the property and is therefore comparatively simple. Where it is for a proposed development the income has to be estimated. The market study should show the range of rents which might be expected, and it is then a matter of judgement on the part of the developer to assume rental rates based on market conditions and the quality and cost of the building he is proposing.

Rental rates vary with the type of building and the basis on which they are charged will also differ. The following gives a general idea of how some rentals are charged:

1. OFFICE BUILDINGS
The rental is based on the rentable floor area. The most usual method of measuring the rentable floor area is described in *Standard Method of Floor Measurement* published by Building Owners and Managers Association International, 224 Michigan Avenue, Chicago, Illinois 60604, U.S.A. Briefly, with a full-floor occupancy it is usually measured to the inside of the exterior walls, with deductions for stairwells, elevators shafts, duct shafts and mechanical spaces only. If the tenant takes only a part of a floor the area is taken to the centre line of the demising partitions and the inside face of exterior and corridor walls, with no deductions for columns. In this case, because the developer does not rent washrooms or corridors, the rental rate will be higher than on a full-floor occupancy. Normally the ground floor will rent at a higher rate than any other floor. The rental rate may be net, with the tenant paying taxes and certain operating costs separately, the lease may contain an escalation clause, or the rent may be fixed and all-inclusive. The variety of ways in which the rental rate can be established is limited only by the developer's imagination and the tenant's acquiescence.

2. APARTMENT BUILDINGS
The rental is on a per suite basis, varying according to the size and quality of the suite, and on the availability of any supplementary facilities.

3. FACTORIES
The rental is usually based on the gross floor area, measured to the outside face of the exterior walls, and the rate is usually on a net net basis with the tenant paying for all operating costs, including taxes.

4. SHOPPING CENTRES
A shopping centre may contain office accommodation, theatres, restaurants and gas stations as well as retail stores and department stores, and the leasing arrangements will differ for each. A common arrangement, except for the office accommodation, is for the tenant to agree to a minimum rent based on the floor area, plus a percentage based on audited receipts. This percentage can vary considerably depending on turnover. The tenant then pays the minimum rent based on the floor area and the agreed percentage on receipts.

OPERATING COSTS

Operating costs can be separated into two types: those which occur annually and those which occur periodically. They can be further subdivided into occupancy, or user, costs and ownership costs. Occupancy costs are those costs which are related to the activities which are being carried on within the building such as the cost of medical staff in a hospital or the costs of running an assembly line in a factory. There may be occasions when occupancy costs

FEASIBILITY STUDIES

and ownership costs need to be separately identified but for the purposes of a feasibility study only the ownership costs usually need to be considered.

Annual ownership operating costs would include;
1. Real estate taxes.
2. Insurance.
3. Cleaning and supplies. This may be done by the developer's own staff or by an outside cleaning organization.
4. Hydro.
5. Fuel.
6. Water.
7. Security.
8. Operating staff.
9. Service contracts. This would include the cost of annual contracts to maintain elevators, air conditioning equipment, etc.
10. Supplies. This would include light bulbs, stationery for the superintendent, etc.
11. Building Management. On large projects this is often handled by the developer's staff in which case it would be included in item 8. Alternatively a professional property management firm may be engaged to collect rents, investigate complaints and make periodic checks of the building. The property management firm may also arrange leasing.
12. Garbage Removal.
13. Snow Removal.
14. Gardening. This would include all outside staff for gardening, pool maintenance, etc.
15. Land rental.
16. Preventive maintenance.
17. Vacancy allowance. If income has been calculated on the basis of a fully-rented building without any allowance for vacancies such an allowance should be made in the annual operating costs.
18. Legal and audit. If a property company has been formed, legal and audit fees will be required, and, in any case, an allowance should be made for preparing lease agreements.
19. External and internal communication equipment. This would include telephones, telex and internal communication systems.
20. Finance costs. This covers payment of interest on loans and will be dealt with in more detail in the next chapter.

Depending upon the terms of the lease some of these costs might be paid directly by the tenants. However, again this provides a useful checklist of annual operating costs which can be expected on a major development.

Periodic ownership operating costs would include:
1. Periodic Redecoration.
2. Periodic Repairs.

3. Potential future alterations. If it is known that alterations will be made at some time in the future, their cost can be estimated and included here, but this is, in fact, a very unlikely item to appear in the feasibility study for a new development.

CHAPTER 8
Financing

Financing is a very important aspect of real estate development. If it is not considered carefully the yield on the developer's investment can be reduced to a much lower rate than it should be, and what had at first sight appeared to be an attractive investment could turn into a financial disaster.

The first consideration in financing is the size of the loan, if any. If the developer wishes to put as little of his own money into the development as possible he will have to take out a large loan, and as a result may have to accept a lower yield on his investment because part of the income from the development will have to be used to pay off the loan. On the other hand, financing a project with his own money may lose some of the leverage which borrowing money can provide. Usually it is advisable to borrow some of the money for a development, but this is not always necessarily the case. If the development is a comparatively modest undertaking such as a small industrial building with a first-class tenant, it is probably better for the developer, if he has the money available, to invest it in the development rather than to borrow.

If the developer decides he needs to borrow money the next consideration is how it will be borrowed. The most common way of doing so is to take out a mortgage on the property but there are other ways. Many major buildings have been financed by mortgage bonds which are sold to the general public through investment dealers and mortgage brokers. One of the chief requirements if mortgage bonds are to be issued is that the development should be backed up by top-class tenants. If an investor is thinking of buying, say, Bell Telephone bonds at 6.5% and he is offered mortgage bonds at 8.5% for a building in which the Bell Telephone Company will be the prime tenant, he would be wise to take the mortgage bonds since they offer an extra 2% and, in effect, they are backed up by the same company. Another alternative to a mortgage is to offer shares in a property company which is to be formed to carry out the development.

Usually however a developer will decide to take out a mortgage and when he is negotiating the loan there are a number of factors which he has to consider to be sure of obtaining the best possible terms. The first is the interest rate. He should check the rates offered by the various mortgage lenders to see which, if any, is the most competitive. The Dominion Interest Act requires that the mortgage agreement shall state the amount of principal and the rate of interest to be compounded yearly or half yearly, not in advance. The effect of this is that although mortgage repayments may be made monthly the interest cannot be compounded more frequently than semi-annually. This will be dealt with in more detail in the next chapter but compounding semi-annually means that interest is calculated every six months at half the stated annual rate of interest.

Some mortgage lenders will discount the loan in order to obtain a better yield. When doing this they charge interest on the principal amount stated in the mortgage agreement, but the actual amount of money advanced to the borrower is less. The effect is that the mortgagor (the borrower) is receiving less money and therefore paying a higher rate of interest than agreed, and the mortgagee (the lender) is achieving a better rate of return.

Every mortgage has an amortization period and a term, and the two are not usually the same. To amortize a loan is to reduce it to zero over a period of time by repaying it in installments. It comes from the French *a mort* meaning "at the point of death" so the loan is in effect being "killed off." The amortization period is the time required to accomplish this, and it, together with the interest rate and the size of the loan will govern the amount of the repayments. It is usual for repayments to be made in equal monthly installments, although other arrangements can be made, with part of the payment going toward interest and the balance going toward repayment of the principal. While the amortization period can be twenty-five years or more the term is usually considerably less, probably no more than five years. The term is the period of time at the end of which the mortgagee can ask for his money back. If the mortgagor cannot repay the loan at the end of the term he has to renew the mortgage or arrange financing elsewhere to pay off the mortgagee, but in either case he is likely to have to pay a different, and probably higher, rate of interest than he was paying on the original mortgage.

At one time it was not uncommon for the amortization period and the term to be the same, but as interest rates climbed and mortgagees found themselves with long-term loans at low rates of interest, they reduced the length of the term in order to take advantage of the higher interest rates. With commercial developments a longer term might be arranged and it is up to the developer to find a mortgage lender who will provide him with money on a term which suits his development.

Another factor which the developer has to consider is the prepayment privileges which depend to a large extent on the mortgage agreement. If the mortgage is an open mortgage the mortgagor can discharge the mortgage at any time by paying off the loan without incurring any penalty. Direct N.H.A.

loans made with Central Mortgage and Housing Corporation are usually of this type. With most other lending institutions there will be some form of penalty if the mortgagor wishes to pay off his mortgage before the end of the term, the amount and type of penalty depending on the requirements of the Dominion Interest Act, the age of the mortgage, the current interest rates and the current demand for mortgage loans.

In some provinces the mortgagee is permitted to restrict his loan to the original mortgagor only. In this case, since the mortgage cannot be transferred, any purchaser of the developer's property would have to arrange his own financing, or the developer would have to take back a second mortgage to enable the sale to go through, either of which could prove to be awkward for the developer at a later date when he wishes to sell the property.

A fairly recent innovation on commercial properties is for the mortgagee to require participation in the development by claiming additional payments if the income is greater than was anticipated. Any developer who enters into this form of agreement should make sure that he obtains something in return for losing the additional income if the development proves to be highly successful.

A mortgage is a charge against real estate, with a first mortgage taking precedence over second and subsequent mortgages, and a second mortgage only becoming a first mortgage when the original first mortgage has been discharged. Any subsequent mortgages would then also move up a step in precedence. If the mortgagor defaults in his repayments on the first mortgage the mortgagee can, after taking the necessary legal steps, foreclose on the property, wipe out all subsequent charges and become the owner of the property with a clear title. This is one of the reasons why second and subsequent mortgages have higher interest rates; the risk is greater because the mortgagor may default on the first mortgage and these mortgagees would then lose their money. A second or subsequent mortgagee can also take action for foreclosure if the mortgagor fails to make payment. However, in this case the mortgagee would become liable for repayment of the loans to all mortgagees who have a higher ranking than he has, while those with a lower ranking would find themselves in the same position as in the previous case: they would lose their money.

The best time to apply for long-term financing is when the building is completed and leasing is well under way. Few lending institutions are keen to offer money when all they can see is a building site and a set of drawings. The following is the basic information which is needed when applying for a mortgage:

1. A copy of the working drawings and specifications of the building if the application is being made before construction starts. These would be reviewed by the lending institution's staff and by outside consultants to determine whether the building is viable.

CONSTRUCTION ESTIMATING AND COSTING

2. A list of proposed tenancies together with the lease agreements showing the rental terms if they are available.

3. A pro forma showing the financial feasibility of the development.

4. Financial statements of the development company or, if it is a company formed specially for the development, of the parent company, accompanied by details of the people who are putting the project together. The last thing the lending institution wants is to become owner of the building and they therefore want to check not only the ability to keep up the mortgage payments but also the willingness to do so.

Provided the developer meets the lending institution's criteria for a loan he will be given a commitment, which is a preliminary contract to provide the loan, and may be asked for a holding deposit of about one or two per cent of the loan, which may or may not be refunded when the mortgage agreement is signed. A non-refundable holding deposit is another way the mortgagee can increase his earnings.

To illustrate how financing can affect the profitability of a development, consider an industrial building of 930 m^2 which can be rented at $12.90 per m^2 net and which costs $100,000 to build. Three means of financing are available: by the developer providing all cash (METHOD A); by means of a mortgage of $75,000 at 9.75% interest with an amortization period of fifteen years (METHOD B); and by means of a mortgage of $60,000 at 10.25% interest with an amortization period of twenty-five years (METHOD C). At first glance it would appear that of the three, METHOD A is probably the best since none of the income has to be paid out on mortgage repayments, while METHOD B is undoubtedly better than METHOD C because it has a lower rate of interest and is paid off quicker. However, a tabulation of the figures shows the following:

	Method A	Method B	Method C
Capital cost	$100 000	$100 000	$100 000
Mortgage	—	75 000	60 000
Developer's equity	100 000	25 000	40 000
Net income:			
930 m^2 @ $12.90	12 000	12 000	12 000
Annual mortgage payments:			
Method A	—		
Method B $75,000 × 12.58		9 435	
Method C $60,000 × 10.94			6 564
Net income	12 000	2 565	5 436

Return on equity

Method A: $\dfrac{12\,000}{100\,000}$ 12%

Method B: $\dfrac{2\,565}{25\,000}$ 10.26%

Method C: $\dfrac{5\,436}{40\,000}$ 13.59%

In these circumstances, therefore, METHOD C gives the best rate of return, followed by METHOD A, and then METHOD B, although a change in the income or the total cost will give quite different results. The multipliers of 12.58 and 10.94 given for calculating the annual mortgage payments are obtained from mortgage amortization tables and show the annual amount to be paid for principal and interest in order to pay off $100 at the given rate of interest for the length of the loan.

This example shows how leverage can work. By borrowing $60,000 at 10.25% interest for a period of twenty-five years the developer can obtain a better return on his investment than he can if he uses his own money. A quick way of telling whether leverage will work in any given circumstance is to compare the rate of return using all cash, sometimes known as the natural constant, with the multiplier for annual mortgage payments for the particular interest rate and length of the loan which is shown in mortgage tables. In the example, the multiplier for METHOD B is 12.58 which is higher than the natural constant of 12.00, an indication that this method of financing is taking more than its share of the net income and will not therefore provide any leverage. On the other hand the multiplier for METHOD C is 10.94, lower than the natural constant, so this method will provide leverage. It should be noted that comparison with the natural constant does not necessarily rank alternative methods of borrowing in order of merit, but only compares a method of borrowing with not borrowing at all to see whether leverage can be obtained.

One further point in connection with leverage is that it can work in reverse so the developer has to be very sure of his figures. If it turned out that instead of obtaining $12.90 per m² in rent he could only obtain $10.50 per m², or if the capital cost were to increase, not only would the return on equity for both METHOD B and METHOD C drop below the return for METHOD A, but also if he found he couldn't rent the building at all he would still have to make his mortgage payments, without having any income to cover them.

The method of calculating the return on equity in this example, while it serves its purpose for the example, does not give a true rate of return, as will be explained in Chapter Ten.

CHAPTER 9

Time And Money

Money has a different value depending on when it it received and for how long it is kept. This is due to two factors: inflation, which tends to reduce its value; and interest, which increases its value over a period of time. This chapter will deal with the second factor, the effects of investing money and thereby earning interest.

Most people have to spend a large part of their income when they receive it but if they have any left over they will want to save it. One way of saving money is to keep it under the mattress although this is not generally recommended because money under the mattress doesn't earn interest. A better way of saving is to invest it since, whether money is spent or saved, everyone would like to receive something worthwhile in return.

Simple Interest

When money is invested the intention is that it should earn interest. Interest is a charge for the use of the money, and the amount of interest earned will depend upon the amount of money invested, the length of time it is invested, and the rate of interest expressed as an annual percentage. It will also depend upon whether the interest is to be calculated as simple interest or compound interest.

With simple interest only the principal, that is the original amount of the investment, will earn interest. If a thousand dollars were invested at ten per cent per annum simple interest for five years, the interest at the end of the first year would be one hundred dollars and, because only the principal earns interest, it would also be a hundred dollars at the end of the second, third, fourth and fifth years for a total of five hundred dollars. The formula for calculating simple interest is:

$$I = Pin$$

where I = the total amount of interest

P = the principal amount
i = the rate of interest (10% = 0.10, 6% = 0.06 etc.)
and n = the time period of the investment.

Thus the interest earned on $1000 at 10% per annum for five years at simple interest is:

$$\$1000 \times 0.10 \times 5 = \$500$$

Compound Interest

Simple interest is fine if it is intended to spend the interest each year and not to reinvest it. However, suppose the money earned in interest is also to be invested so that it, in turn, can earn interest. If simple interest were the only way of investing money it would mean that the interest earned each year would have to be withdrawn so that it could be redeposited as principal to earn interest. Fortunately this doesn't have to be done because when money is deposited at compound interest the interest is automatically added to the principal each year and the two combined will earn further interest.

If the thousand dollars were invested at ten per cent compound interest for five years it would be worth one thousand one hundred dollars at the end of the first year. This amount will earn interest of one hundred and ten dollars by the end of the second year to give a total of one thousand two hundred and ten dollars, and this will continue until at the end of the fifth year the total amount of interest will be $610.51, and the original investment of a thousand dollars will have accumulated to $1610.51. This is known as the terminal value of the investment, its amount depending on the amount of the investment, the rate of interest, the length of time of the investment and when it is made. The terminal value will not be the same if the investment is delayed for six months since there will then be a period of six months when the money is not earning interest.

Besides finding the terminal value of an investment there are other calculations needed by an investor to find out the effect of time on money, and because they can become quite complicated, discount tables, or interest and annuity tables for real estate appraisers, which are almost identical, have been published. The tables give the amounts for the various functions based on one dollar at differing rates of interest and over different time periods, thus eliminating the need to calculate the figures each time they are needed, aside from multiplying the figure in the tables by the actual sums being used. Computer programs are also available to do the calculations, as are electronic calculators with built-in financial functions or with an exponential function. In fact an electronic calculator with financial functions or with an exponential function can, with practice, be quicker than looking up the figures in the tables, and it has the added advantage that it is not limited to those interest rates which are published in the tables.

Future Worth of One Dollar (A)

THE FUTURE WORTH OF ONE DOLLAR, also known as THE AMOUNT OF ONE DOLLAR, is the terminal value of one dollar invested now at a given rate of compound interest over a given period of time, and is the basis for all other calculations. It is shown graphically in Figure 6.

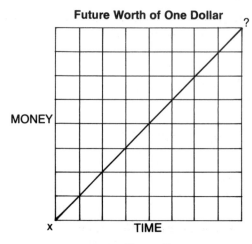

Figure 6

If an amount of one is to be invested at compound interest at a rate 'i' per annum, it will have amounted to 1+i at the end of the first year. By the end of the second year, the interest earned during that year will be (1+i)i, and this together with the amount invested at the beginning of the second year (that is the original principal plus interest earned during the first year) will give a terminal value of:

$$(1+i) + (1+i)i$$
$$= 1 + i + i + i^2$$
$$= 1 + 2i + i^2$$
$$= (1+i)^2$$

Interest earned during the third year will be $(1+2i+i^2)i$, so the terminal value at the end of the third year will be:

$$(1 + 2i + i^2) + (1 + 2i + i^2)i$$
$$= 1 + 2i + i^2 + i + 2i^2 + i^3$$
$$= 1 + 3i + 3i^2 + i^3$$
$$= (1 + i)^3$$

By extention it can be seen that the formula for this calculation is:

$$A = (1 + i)^n$$

where A = the future worth of one (sometimes written as S^n)
i = the rate of interest
and n = the time period.

It should be noted that the rate of interest and the time period should correspond. If the rate of interest is an annual rate then the time period should be in years, and if the time period is in months then the rate of interest should be a monthly rate. This will be raised again when compounding periods are considered.

Example: How much will $150 amount to in 21 years if it is invested at 6% per annum compound interest?

Answer: $150 × (1 + 0.06)^{21}
        = $150 × 3.3996
        = $509.94

Present Worth of One Dollar (V)

THE PRESENT WORTH OF ONE DOLLAR is the amount which must be invested now at a given rate of compound interest to accumulate to one dollar over a given period of time, and is shown graphically in Figure 7.

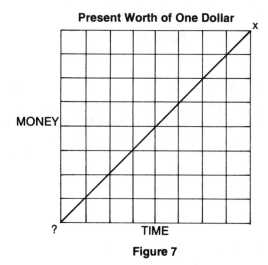

Figure 7

This, like the FUTURE WORTH OF ONE DOLLAR, is a straight compound interest calculation, but whereas with the FUTURE WORTH OF ONE DOLLAR, the investor starts with a dollar and finishes up with his dollar plus compound interest, with the PRESENT WORTH OF ONE DOLLAR he finishes up with a dollar and needs to know how much he must start with for it to accumulate to a dollar over a given period of time at compound interest.

CONSTRUCTION ESTIMATING AND COSTING

A little thought will show that the PRESENT WORTH OF ONE DOLLAR is the reciprocal of the FUTURE WORTH OF ONE DOLLAR, so the formula is:

$$V = \frac{1}{(1 + i)^n} \text{ or } \frac{1}{A}$$

where V = the present worth of one (sometimes written as V^n)

The present worth is also sometimes known as the present value or deferred value and the process of allowing for the fact that a sum will not be received until some time in the future is known as deferring or discounting that sum.

Example: How much should be invested now to give $1000 in 12 years at 6.5% per annum compound interest?

Answer: $1000 \times \dfrac{1}{(1 + 0.065)^{12}}$
= $1\,000 \times 0.4697$
= $469.70

Future Worth of One Dollar per Annum (FV)

THE FUTURE WORTH OF ONE DOLLAR PER ANNUM, also known as the AMOUNT OF ONE DOLLAR PER ANNUM is the terminal value of one dollar a year invested at a given rate of compound interest over a given period of time, and is shown graphically in Figure 8.

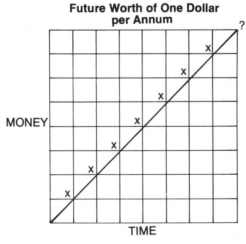

Figure 8

An annuity is a series of equal payments made at regular intervals and the future value of an annuity can be calculated using the formula for the FUTURE WORTH OF ONE DOLLAR PER ANNUM, although with many annuities the payments are not made on an annual basis but on a monthly, quarterly or some other basis, in which case the interest rate and time periods must be adjusted accordingly.

The formula for calculating the future worth will vary depending on whether the money is invested at the beginning or the end of each period. When the money is invested at the end of the period, as for example mortgage repayments which start a month after the loan is made, it is sometimes known as an ordinary annuity. The formula for calculating the FUTURE WORTH OF ONE DOLLAR PER ANNUM when the investment is made at the *end* of the period is:

$$FV = \frac{(1+i)^n - 1}{i} \quad \text{or} \quad \frac{A-1}{i}$$

where FV = the future worth of one per period (sometimes written as $S_{\overline{n}|}$)

When the money is invested at the beginning of the period, as for example insurance premiums or rents which are paid in advance, it is sometimes known as an annuity due. The formula for calculating the FUTURE WORTH OF ONE DOLLAR PER ANNUM when the investment is made at the *beginning* of the period is:

$$FV = \frac{(1+i)^{n+1} - 1}{i} - 1$$

Example: How much will $150 a year amount to in 21 years if invested at 6% per annum compound interest (a) with the first installment being made immediately, and (b) with the first installment being made in a year's time?

Answer: (a) $150 × $\frac{1.06^{22} - 1}{0.06} - 1$

= $150 × 42.3923
= $6358.85

(b) $150 × $\frac{1.06^{21} - 1}{0.06}$

= $150 × 39.9927
= $5998.91

Annual Sinking Fund (SF)

THE ANNUAL SINKING FUND factor is the amount which must be invested each year to accumulate to one dollar at a given rate of compound interest over a given period of time. It is shown graphically in Figure 9.

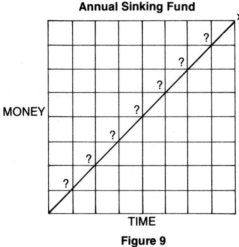

Figure 9

When a corporation or government issues bonds it knows that it will have to repay the loan when the bonds mature. In order to ensure that it can meet the repayment at maturity it creates (or should create) a separate fund into which it makes equal periodic deposits which, with compound interest, will equal the amount of the loan by the time the loan has to be repaid. This is called a sinking fund because its purpose is to sink, or shrink, the debt. The interest earned on the sinking fund is not necessarily at the same rate as the interest paid on the bonds. A similar circumstance arises with a developer who is putting up a new building. A building, unlike land, is a wasting asset because eventually it will come to the end of its life and have to be demolished. If the developer doesn't put aside part of his income and deposit it in a sinking fund he will one day find himself with a piece of land, with no building on it, and insufficient capital to construct a new building. Whether, in fact, he does set up a sinking fund is a decision the developer has to make and he may decide not to, but nevertheless a sinking fund is usually incorporated in the calculation of the present value of a development.

Sinking fund payments are made at the end of each period so the formula is the reciprocal of the FUTURE WORTH OF ONE DOLLAR PER ANNUM when the investment is made at the end of the period, thus:

$$SF = \frac{i}{(1+i)^n - 1} \quad \text{or} \quad \frac{i}{A-1}$$

where SF = the sinking fund factor (sometimes written as $1/S_{n\rceil}$)

Example: How much should be invested at the end of each year to give $5000 in 10 years' time at 6% compound interest?

Answer: $\$5000 \times \dfrac{0.06}{1.06^{10} - 1}$

$= \$5000 \times 0.07587$
$= \$379.35$

Present Worth of One Dollar per Annum (YP)

THE PRESENT WORTH OF ONE DOLLAR PER ANNUM, which is shown graphically in Figure 10, will depend upon whether the income is to be earned perpetually or whether there is a limited life to it, and if the latter is the case whether the income starts immediately or starts at the end of the first period.

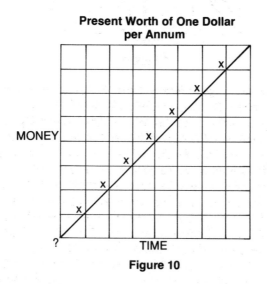

Figure 10

If an investment of $100,000 is made in, say, a piece of land, which can be said to provide a perpetual income because it is not a wasting asset, and the investor expects to receive a 10% return on his investment, he can expect to receive an annual income of:

$$\dfrac{10}{100} \times \$100\,000 = \$10\,000$$

This is a simple calculation, based on simple interest, showing the income which the investor expects to receive on his investment but making no

CONSTRUCTION ESTIMATING AND COSTING

statement about what he will do with the income. Reversing the process, if the investor wishes to invest a sum of money which will give him a 10% return amounting to $10,000 a year, the amount he needs to invest is:

$$\frac{100}{10} \times \$10\,000 = \$100\,000.$$

The multiplier which has to be applied to the annual income to give the capital investment is known as the PRESENT WORTH OF ONE DOLLAR PER ANNUM IN PERPETUITY. In the United Kingdom it is also known as YEARS' PURCHASE. It can be calculated very easily by dividing one hundred by the rate of interest, thus:

If the rate of interest is 5%, YP = $\frac{100}{5}$ or 20

If the rate of interest is 4%, YP = $\frac{100}{4}$ or 25

If the rate of interest is 6%, YP = $\frac{100}{6}$ or 16.67

Note that these calculations apply only when the income is perpetual.

If the investor purchases an income for a limited time, he is in quite a different position from the purchaser of a perpetual income. Whereas the purchaser of a perpetual income can expect to receive the income for as long as he lives or until he sells his investment, the purchaser of an income for a limited time will receive the income only for that period, and he will then lose not only his income but also his initial investment. This is the case if he purchases an income-producing building because buildings have a limited life. In order to recapture his initial investment he needs to include a sinking fund in his calculation, and this is what happens in the formula for the PRESENT WORTH OF ONE DOLLAR PER ANNUM, also known as the INWOOD ANNUITY FACTOR, or YEARS' PURCHASE WITH A LIMITED TERM. This formula gives the amount which must be invested now to give an income of one dollar a year together with a sinking fund to replace the investment, at given rates of simple interest for the investment and compound interest for the sinking fund, over a given period of time.

In most circumstances the income starts at the end of the first period, not immediately, and it has to equal the amount of the interest plus the sinking fund, that is $i + SF$. Since the capital investment (P) equals the income times the YEARS' PURCHASE, then:

$$P = (i + SF) \times YP$$

and $YP = \dfrac{P}{(i + SF)}$

In finding the PRESENT WORTH OF ONE DOLLAR PER ANNUM where the capital investment is one dollar, the formula becomes:

$$YP = \frac{1}{i + SF}$$

or

$$\frac{1}{i + \frac{i}{(1+i)^n - 1}}$$

Where YP = the present worth of one per period (sometimes written as a $\overline{n\rceil}$)

This is the most common formula for calculating the PRESENT WORTH OF ONE DOLLAR PER ANNUM.

It is not usual to start receiving an income immediately on an investment, but in the event that this does occur the formula becomes:

$$YP = \left(\frac{1}{i + SF}\right)(1 + i) \text{ or}$$

$$\left[\frac{1}{i + \frac{i}{(1+i)^n - 1}}\right](1 + i)$$

Note that in these formulae there are in fact two rates of interest involved, one known as the remunerative rate which is used to provide the income, and a second known as the accumulative rate which is used to provide the sinking fund to replace the capital when the investment ceases. For this reason the published discount tables very often give two separate tables for the PRESENT WORTH OF ONE DOLLAR PER ANNUM, a single rate table and a dual rate table. A single rate table gives figures when the remunerative interest rate and the accumulative interest rate are the same, while the dual rate table gives the figures when the two rates are different. In the past the accumulative rate for the sinking fund was lower than the remunerative rate because it was assumed that the sinking fund investment would be in low-risk bonds, or in a sinking fund policy taken out with an insurance company, either of which would bear low interest rates. It is usual these days, however, to assume that both rates will be the same, and when referring to published tables it is the single rate table which is normally used.

Example: How much should be invested in a building which will be demolished in ten years' time if an income of $20 000 a year can be expected from it, and a reasonable discount rate is 6%?

Answer: $20\,000 \times \dfrac{1}{0.06 + \dfrac{0.06}{1.06^{10} - 1}}$

$= \$20\,000 \times 7.3601$

$= \$147\,202.$

Annual Equivalent Worth

The ANNUAL EQUIVALENT WORTH, also known as the PARTIAL PAYMENT FACTOR, gives the annual income which will be earned on an investment of one dollar at a given rate of interest, and is shown graphically in Figure 11.

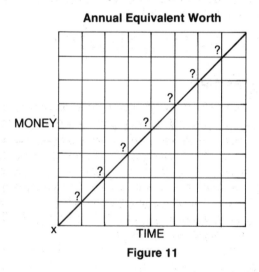

Figure 11

Because it is the reciprocal of the PRESENT WORTH OF ONE DOLLAR PER ANNUM it will have the same variations as are found with that function. If the income is to be received in perpetuity the income will be P × i as has already been shown. If, however, it is to be received for a limited time so that the sinking fund has to be taken into consideration and income is to start at the end of the first period, the formula for the ANNUAL EQUIVALENT WORTH, based on an investment of one dollar, will be:

$$\text{AEW} = i + \text{SF or } i + \dfrac{i}{(1 + i)^n - 1}$$

where AEW = the annual equivalent worth of one (sometimes written as $1/a_{\overline{n}|}$)

This formula can also be used for calculating the payments required to amortize a mortgage, and is the most usual for calculating the ANNUAL EQUIVALENT WORTH.

Example: If $150 is invested at 6% for 21 years, what will the annual income be assuming that the principal will be lost at the end of the investment?

Answer: $$\$150 \times 0.06 + \frac{0.06}{1.06^{21} - 1}$$
$$= \$150 \times 0.085$$
$$= \$ 12.75$$

If the income is to be received immediately rather than starting at the end of the first period, the formula will be:

$$AEW = (i + SF)(1 + i) \text{ or } \left(i + \frac{i}{(1 + i)^n - 1}\right)(1 + i)$$

Compounding Periods

All the examples given so far in this chapter have assumed that compounding will take place annually, although reference was made earlier to the fact that it might take place more frequently. When compounding takes place annually the assumption is that the principal will remain invested for a year, the interest will then be added to it, the two together will accrue interest at the end of the second year, and so on year by year with interest being added on as an annual event. With annual compounding therefore, "i" in the formulae will always be the annual interest rate, and "n" will always be the number of years over which the investment takes place.

Interest is not always compounded annually. It can be compounded semi-annually, quarterly or weekly, and presumably there is no reason why it shouldn't be compounded daily, hourly or even more frequently. If, for example, interest is to be compounded quarterly, interest is calculated at the end of the first quarter, added to the principal, and the two combined are carried forward for interest to be calculated at the end of the second quarter, and so on, for the life of the investment. In other words interest is added, not as an annual event but as a quarterly event. This does not affect the formulae but means that "i" is no longer an annual rate and "n" is no longer in years. To illustrate what happens when compounding is done more frequently than annually, suppose one hundred dollars is to be invested at eight per cent for one year. The future value at various compounding periods will be as follows:

Annual: $100 × (1 + 0.08)^1$ = $108.00
Semi-annually: $100 × (1 + 0.04)^2$ = $108.16
Quarterly: $100 × (1 + 0.02)^4$ = $108.24
Monthly: $100 × (1 + 0.00667)^{12}$ = $108.30

When compounding more frequently than annually "n" becomes the number of compounding periods rather than the number of years, and "i" is the annual interest rate divided by the number of compounding periods in a year. This example also shows that by compounding more frequently than annually the investor will receive a better return on his money. In fact, although the nominal or stated rate is eight per cent in each calculation, by compounding monthly the actual rate has been increased to 8.30%. The actual rate is known as the effective rate. When compounding annually the nominal and the effective rate will be the same, but when compounding more frequently the effective rate will always be higher than the nominal rate.

Compounding more frequently than annually is of particular interest when calculating mortgage repayments.

Example: What will the monthly payments be on a mortgage of $30,000 with an amortization period of 25 years at an annual interest rate of 11%, compounded monthly?

Answer: $AEW = i + \dfrac{i}{(1 + i)^n - 1}$

$= 0.009167 + \dfrac{0.009167}{1.00967^{300} - 1}$

$= 0.009801$

Monthly payments = $30\,000 × 0.009801$
$= \$294.03$

Note that the interest rate is shown as a monthly rate $\left(\dfrac{0.11}{12} = 0.009167\right)$ and "n" is the number of months (25 × 12 = 300).

In Canada, as was mentioned in the last chapter, the Dominion Interest Act stipulates that interest on a mortgage cannot be compounded more frequently than semi-annually, although payments may be made monthly. This means that although the principal is being reduced each month as part of the mortgage payment, the compounding can take place only every six months, which introduces an added complication to the calculation. It is overcome by adjusting the nominal interest rate to an effective monthly rate by means of a further formula which is:

Effective monthly rate $= \left(1 + \dfrac{i}{2}\right)^{1/6} - 1$

Example: What will the monthly payments be in the previous example if interest is to be compounded semi-annually?

TIME AND MONEY

Answer: Effective monthly rate

$$= \left(1 + \frac{0.11}{2}\right)^{1/6} - 1$$
$$= 0.008963$$

$$AEW = i + \frac{i}{(1 + i)^n - 1}$$
$$= 0.008963 + \frac{0.008963}{1.008963^{300} - 1}$$
$$= 0.009625$$

Monthly payments = $30\,000 \times 0.009625$
$$= \$288.75$$

Another occasion when compounding more frequently than annually is required is when the cost of interim financing has to be calculated.

Example: What will the cost of financing be on a building expected to cost $8 000 000 with a four-year construction period if the loan is to bear interest at 12% per annum compounded semi-annually with the following schedule of requirements:

First six months	$1 500 000
Second six months	$3 000 000
Third six months	$2 500 000
Fourth six months	$1 000 000

Answer:
First draw	$1 500 000 × $1.06^{3\,1/2}$	= $1 839 339
Second draw	$3 000 000 × $1.06^{2\,1/2}$	= 3 470 451
Third draw	$2 500 000 × $1.06^{1\,1/2}$	= 2 728 342
Final draw	$1 000 000 × $1.06^{1/2}$	= 1 029 563
		$9 067 695
	Less Principal	8 000 000
	Cost of financing	$ 1 067 695

These calculations are onerous if they have to be done by hand but fortunately tables are readily available for Canadian amortization schedules. An electronic calculator with exponential or financial functions makes the calculations comparatively easy if they have to be worked out without the aid of tables.

CHAPTER 10

Yield Analysis

Yield

Yield is the return an investor expects to make on his investment, usually expressed as a percentage. For many investments the calculation of the yield is fairly simple; a mortgagee for instance knows that, subject to the mortgagor defaulting, his yield will be the stipulated interest rate on the unpaid balance of the mortgage. Similarly, a bond purchased at face value will give a yield which is the same as the nominal rate of the bond. However, if the bond is bought at anything other than the face value, the yield will be different from the nominal rate. As an example, a bond with a face value of $1000 and selling at face value will yield $50 a year if its nominal rate is 5%, so the yield and the nominal rate are the same. But if the bond can be bought for $850 the income will still be $50 a year and the yield will be:

$$\frac{5}{850} \times 1000 = 5.88\%$$

or $\frac{50}{850} \times 100 = 5.88\%$

This simple example shows that a comparison of incomes receivable from investments should be made on the yields, not on the nominal interest rates. It also shows that a rise in the price of a security will result in a drop in the yield, while a drop in the price will increase the yield, assuming the yield is based solely on the income from the interest.

For the real estate investor the calculation of the yield becomes more complicated, since it has to be made on a number of assumptions. In some instances the yield will be based solely on the income to be derived from the property, in which case assumptions have to be made about the income, whether it will remain constant throughout the life of the investment, whether it is likely to increase over the years or whether it may decline. In other instances the investor may assume that he will sell the property after a few years, in which case he has to make additional assumptions about the selling price, whether it will be the same as his cost, whether it will be more than his cost or whether it will result in a capital loss.

Each investor will have his own objectives and priorities when making investment decisions. Generally they will be to maximize cash flow and improve capital gain prospects, but they might also include considerations of prestige, amenities and tax implications. To the extent that an investor may be prepared to accept higher or lower yields, the criteria by which investments are usually judged are as follows:

1. SECURITY OF INCOME

The investor likes to be assured that he will receive a regular income from his investment, but even more importantly he would like to be assured that the income will keep pace with inflation. An adequate income now will prove to be quite inadequate in the future if the purchasing power of the dollar declines rapidly, and an investor who can be assured that his income is secure and will increase at the same, or at a better rate than the decline in the value of money is likely to be prepared to accept a lower yield than if there is no such assurance.

2. SECURITY OF CAPITAL

As with security of income, the investor likes to be assured that the value of his investment will keep pace with inflation. If he buys a piece of property which appreciates in value by fifty per cent in five years when he sells it, it might be said that he has made a gain of fifty per cent; but if all other commodities have also appreciated by fifty percent, then in real terms he has made no gain at all and the value of his asset is the same as when he bought it. In the days when there was no tax on capital gains many investors were not too concerned about the annual income from the investment and would be prepared to accept a lower yield if it could be shown that there would be an appreciation of capital over the years.

3. LIQUIDITY

This refers to the ease with which an asset can be converted into cash, and can also encompass the acceptability of the asset as collateral for a loan. Real estate is not as liquid as, say, stocks traded on the stock market, and an investor would require a higher yield on a real estate investment than he would if his assets were in say, stocks and bonds.

4. TRANSFER COSTS

The acceptable yield would depend to some extent on the costs to invest the capital and to subsequently convert it back into cash.

5. MANAGEMENT

Real estate investments require management of the investment portfolio, more so than investments in government bonds, and the investor would therefore expect a higher yield. Note that in this context management does not refer to the day-to-day management of the property.

6. LEGISLATION

Most types of investment are affected to some degree by government legislation but none to the same extent as real estate. An investor whose investment may be affected by rent controls or some other form of legislation restricting the income he can expect will require a higher yield.

Capitalization Methods

In chapter seven it was stated that a feasibility study attempts to show that the value of a real estate development at least equals its cost. The capital cost of a development is comparatively easy to determine, but its value, since it refers to present value, is not so easy to find because it requires assumptions to be made about future incomes and expenses which then have to be discounted using the formulae described in the last chapter. Discounting also needs interest rates, known as capitalization rates, and a major difficulty is to find that capitalization rate which is also the yield rate.

Table 7 shows the figures for a simple development in which the developer will be providing all the capital.

Table 7

Capital Costs:	Land	$100 000
	Building and development costs	900 000
	Total	$1 000 000
Gross Annual Income:		$170 000
Gross Annual Expenses: Operating and maintenance costs		$ 70 000
Net Annual Income:		$100 000

YIELD ANALYSIS

Example 1: What is the capitalization rate for the development shown in Table 7?

Answer: The capitalization rate will be the net annual income divided by the investment, multiplied by one hundred:

$$\frac{\$100\ 000}{\$1\ 000\ 000} \times 100 = 10\%$$

This example shows that the capitalization rate is ten per cent, but is it also the yield rate? Will the developer receive a ten per cent return on his investment of a million dollars? The answer to this question is found by capitalizing the future incomes from the property to find their present value, and if the present value is equal to the cost, then the capitalization rate is, in fact, the yield rate. Capitalization uses the formulae described in the last chapter. For future annual income the formula for the PRESENT WORTH OF ONE DOLLAR PER ANNUM (YP) is required, and for a future lump sum income the formula for the PRESENT WORTH OF ONE DOLLAR (V) is required. In the next examples a capitalization rate of ten per cent will be used since this is the rate found in Example 1.

Example 2: If the developer expects to sell the development after ten years for $1 000 000 will his yield be 10%?

Answer: Income $100 000 × YP, 10 years @ 10%
= $100 000 × 6.144567106   = $614 457

Sale $1 000 000 × V, 10 years @ 10%
= $1 000 000 × 0.385543289   =   385 543

Total Present Value $1 000 000

The total present value equals the capital cost, so the yield is 10%.

Example 3: If the developer expects to sell the development after twenty years for $1 000 000 will his yield still be 10%?

Answer: Income $100 000 × YP, 20 years @ 10%
= $100 000 × 8.51356372   = $851 356

Sale $1 000 000 × V, 20 years @ 10%
= $1 000 000 × 0.148643628   =   148 644

Total Present Value $1 000 000

The total present value still equals the capital cost, so the yield is still 10%.

CONSTRUCTION ESTIMATING AND COSTING

Example 4: Suppose the developer expects to sell the development after ten years for $1 500 000 will his yield still be 10%?

Answer: Income $100 000 × YP, 10 years @ 10%
                  = $100 000 × 6.144567106   = $614 457

         Sale $1 500 000 × V, 10 years @ 10%
                  = $1 500 000 × 0.385543289   =   578 315
         Total Present Value $1 192 772

The total present value is now more than the cost and the yield will be more than 10% because of the capital gain made on the sale.

In the next example the land, being a non-wasting asset, has been separated from the building and development costs when calculating the present value, a method frequently used when valuing a property.

Example 5: If the developer decides to hold the property for 40 years, at which time he expects it to be demolished, will his yield still be 10%?

Answer: Net annual income $ 100 000
         Return required on land:
            $100 000 × AEW in perpetuity @ 10%
            = $100 000 × 0.10   =   10 000
         Balance attributable to the building $ 90 000

         Income $90 000 × YP, 40 years @ 10%
                  = $90 000 × 9.779050719   = $ 880 115
         Present value of the land   100 000
         Total Present Value $980 115

The total present value is less than the cost so the yield will be less than 10%.

The same present value will in fact also be found if the approach taken in the previous examples is applied:

         Income $100 000 × YP, 40 years @ 10%
                  = $100 000 × 9.779050719   = $ 977 905
         Residual land value $100 000 × V, 40 years @ 10%
                  = $100 000 × 0.022094928   =   2 209
         Total Present Value $ 980 114

Only if the land were to increase in value so that it became worth one million dollars (the amount of the initial investment) in forty years time would the present value equal the cost.

YIELD ANALYSIS

These examples show that, while the calculation given in Example 1 may give an indication of the viability of a project, it does not necessarily give the yield that can be expected. Example 2 and Example 3 show that time is not a factor in deriving the present value to check the yield rate. What is important is the income and the residual value of the property at the end of the investment. The capitalization rate will equal the yield using the method shown in Example 1 only if:

(a) The income remains constant throughout the life of the investment, and

(b) The developer can recover exactly the amount of his capital cost at the end of the investment.

Since these two requirements are seldom met, except in the case of an investment such as a mortgage loan where the mortgagee receives interest payments regularly throughout the term of the mortgage and recovers exactly the amount of the principal at the termination of the mortgage, it is rare that the capitalization rate will equal the yield.

Another rarity is that the developer will use all his own capital to finance the project; it is much more likely that he will take out a mortgage. Table 8 shows the same development as Table 7 but this time the developer has taken a mortgage of $750 000 with interest at 11% and an amortization period of twenty-five years.

Example 6: What is the capitalization rate for the development shown in Table 8?

Answer: The capitalization rate will be the net annual income divided by the owner's equity, multiplied by one hundred:

$$\frac{\$ 13\ 372}{250\ 000} \times 100 = 5.35\%$$

Table 8

Capital Costs:	Land	$100 000
	Building and development costs	900 000
	Total	$1 000 000
Gross Annual Income		$ 170 000
Gross Annual Expenses:	Operating and maintenance costs	70 000
	Mortgage repayments	86 628
	Total	$ 156 628
Net Annual Income		$ 13 372
Owner's equity:	Total capital costs	$1 000 000
	Less Mortgage	750 000
	Total	250 000

Is this also the yield rate? The answer to this, as with the calculations without a mortgage, is that it will be the yield rate only under certain conditions.

The capitalization rate will equal the yield only if:
(a) The income remains constant throughout the life of the investment, and
(b) The developer receives exactly the amount of his equity when the development is sold.

Example 7: If the developer expects to sell the development after ten years, and expects to receive his equity back at that time, will his yield be 5.35%?

Answer: Income $ 13 372 × YP, 10 years @ 5.35%
                  = $ 13 372 × 7.592593799   = $101 528
         Sale $250 000 × V, 10 years @ 5.35%
                  = $250 000 × 0.593887343   =   148 472
         Total Present Value $250 000

The present value equals his equity so the yield is 5.35%.

This is rather a ridiculous situation because every time the developer makes a mortgage payment he pays back a portion of the principal, thus building up his equity in the development. The only way, therefore, in which his equity could remain constant would be if the property were sold at a price equal to the original cost less exactly the amount of the mortgage principal repayments he has made, a highly unlikely occurrence.

There are other ways in which the figures can be manipulated, but none of them necessarily give the yield because they depend on a capitalization rate which may not be the same as the yield rate. For example, as an extension to the previous examples the present value might be divided by the life of the project to give the average annual income and the result expressed as a percentage of the capital outlay. Doing this for the various examples already given has the following results:

Example 2: $\dfrac{\$1\,000\,000}{10} = \$100\,000$

$\dfrac{\$100\,000}{\$1\,000\,000} \times 100 = \underline{\underline{10\%}}$

Example 3: $\dfrac{\$1\,000\,000}{20} = \$50\,000$

$\dfrac{\$50\,000}{\$1\,000\,000} \times 100 = \underline{\underline{5\%}}$

Example 4: $\dfrac{\$1\ 192\ 772}{10} = \$119\ 277$

$\dfrac{\$119\ 277}{\$1\ 000\ 000} \times 100 = \underline{\underline{11.93\%}}$

Example 5: $\dfrac{\$980\ 115}{40} = \$24\ 503$

$\dfrac{\$24\ 503}{\$1\ 000\ 000} \times 100 = \underline{\underline{2.45\%}}$

Example 7: $\dfrac{\$250\ 000}{10} = \$25\ 000$

$\dfrac{\$25\ 000}{\$250\ 000} \times 100 = \underline{\underline{10\%}}$

These results show an interesting variation but are not very informative and certainly don't give any indication of the yield.

Capitalization methods may give an indication of the viability of a project but they don't necessarily give the yield. The calculations just shown can also be used to show the ranking if several alternative development proposals are under consideration, but the NET PRESENT VALUE method is better for this purpose.

Net Present Value

Examples 1 and 6 show the capitalization method, while Examples 2, 3, 4, 5 and 7 were used to show whether and under what circumstances the capitalization rates thus calculated were also the yield. In Examples 4 and 5 the capitalization rate was in fact shown not to be the yield, and the actual yield is still unknown. Examples 2, 3, 4, 5 and 7 are a form of the NET PRESENT VALUE approach in which discounted cash flow analysis is used to show whether a development will provide the yield which the developer considers appropriate. In the NET PRESENT VALUE approach the developer knows what yield he would like to obtain and the minimum yield is therefore given. The yield rate is then used as the capitalization rate, and the answer will be a simple yes or no to the question, will the development provide the required yield?

Example 8: Will the developer obtain a yield of at least 12% for the development shown in Table 7 if he expects to sell it after ten years for $1 500 000?

Answer: Example 4 showed that the development will yield more than 10% under these conditions, and the problem now is whether it will yield as much as 12%.

```
Income    $100 000 × YP, 10 years @ 12%
        = $100 000 × 5.650223028       =      $565 022
Sale      $1 500 000 × V, 10 years @ 12%
        = $1 500 000 × 0.321973237     =       482 960
Total Present Value                           $1 047 982
```

The developer will obtain a yield of at least 12% because the present value exceeds the cost.

Very often the net annual income will not be a constant amount but will fluctuate year by year. A cash flow schedule will then have to be prepared showing the net income year by year, and each year's income must be discounted at the anticipated yield rate to give a total present value. Hence the term discounted cash flow analysis.

Example 9: A developer has a development which has a total capital cost of $900 000 and which he expects to sell for $1 250 000 after ten years. The anticipated net income is as follows:

Year	Income
1	$80 000
2	$90 000
3	$95 000
4	$97 000
5	$100 000
6	$98 000
7	$100 000
8	$100 000
9	$95 000
10	$95 000

Will he obtain a yield of at least 12%?

Answer: The discounted cash flow will be:

Year	Net Cash Flow	× V@ 12%	=	Present Value
1	80 000	× 0.892857143	=	71 429
2	90 000	× 0.797193878	=	71 747
3	95 000	× 0.711780248	=	67 619
4	97 000	× 0.635518078	=	61 645
5	100 000	× 0.567426856	=	56 743
6	98 000	× 0.506631121	=	49 650
7	100 000	× 0.452349215	=	45 235
8	100 000	× 0.403883228	=	40 388
9	95 000	× 0.360610025	=	34 258
10	95 000	× 0.321973237	=	30 587
10 (Sale)	1 250 000	× 0.321973237	=	402 467
Total Present Value				$ 931 768

The development will return a yield of at least 12% because the present value of the income ($931 768) exceeds the cost ($900 000).

YIELD ANALYSIS

The developments shown in Examples 8 and 9 will both return a yield of at least twelve per cent to the developer but if he has a choice between the two how does he decide which is the better investment? There are two ways in which this can be done, the first being to calculate a present value ratio for each so that a comparison can be made. The present value ratio is the total present value divided by the cost. For Examples 8 and 9 the present value ratios will be:

Example 8: $\dfrac{1\ 047\ 982}{1\ 000\ 000} = 1.047982$

Example 9: $\dfrac{931\ 768}{9900\ 000} = 1.035298$

Both investments will yield at least twelve per cent because the present value ratio in both instances exceeds one, but Example 8 has a higher ratio and is therefore a better investment.

The second way of comparing two investments is by marginal analysis. This compares the benefits and costs of each investment and shows the marginal gain (or loss) of one investment over the other. For Examples 8 and 9 the marginal analysis will be:

	Example 8	−	Example 9	=	Margin
Benefit	$1 047 982	−	$931 768	=	$116 214
Cost	$1 000 000	−	$900 000	=	$100 000
Gain	$ 47 982	−	$ 31 768	=	$ 16 214

This shows that for an additional expenditure of $100 000 Example 8 makes a marginal benefit of $116 214 therefore providing an additional "profit" of $16 214, which makes it the better investment.

Comparing the present value ratios and making a marginal analysis will always give the same ranking order. However, they assume that the yield on the difference in capital costs between the two developments will be the same. For example, the difference in capital cost between Examples 8 and 9 is one hundred thousand dollars and it is assumed that the hundred thousand dollars can also be invested at no less than twelve per cent. If, however, it could be invested elsewhere at a considerably higher rate, then it might be preferable to choose the development shown in Example 9 and invest the hundred thousand dollars elsewhere at the higher rate.

A similar difficulty arises with respect to differences in the length of time over which an investment takes place.

Example 10: Will the developer obtain a yield of at least 12% for the development shown in Table 7 if he expects to sell the development after five years for $1 250 000, and what will the present value ratio be?

Answer: Income $100 000 × YP, 5 years @ 12%
= $100 000 × 3.604776203   = $360 478

Sale $1 250 000 × V, 5 years @ 12%
       = $1 250 000 × 0.567426856   =   709 284
Total Present Value $1 069 762

The developer will obtain a yield of at least 12%
The present value ratio will be
$$\frac{1\ 069\ 762}{1\ 000\ 000} = 1.069762$$

This is a better present value ratio than either of the two previous examples, but the yield has been earned over five years whereas the other two were earned over ten years, so can they really be compared? The answer is that they can be compared provided the assumption is made that the developer reinvests the capital he made in Example 10 for a further five years in an investment which also yields at least twelve per cent. This is known as a reinvestment assumption. If he reinvests at any other rate than twelve per cent the present value ratio of Example 10 cannot be compared with the present value ratios of Examples 8 and 9.

Example 11:   If the developer expects to be able to obtain a yield of only 10% for five years upon reinvestment at the conclusion of the investment described in Example 10, will he still obtain a yield of at least 12% over the next ten years, and what will the present value ratio be?
Answer: Total present value, as shown in Example 10: $1 069 762
              Future value in five years' time:
              $1 069 762 × A, 5 years @ 12%
              = $1 069 762 × 1.762341683   = $1 885 286

An investment of $1 885 286 in five years' time, made over five years with a yield of ten per cent will give a future value ten years from now of:
       $1 885 286 × A, 5 years @ 10%
       =$1 885 286 × 1.61051   = $3 036 272

The present value of a sum of $3 036 272 in ten years from now will be:
       $3 036 272 × V, 10 years @ 12%
       = $3 036 272 × 0.321973237   = $ 977 598

The developer will obtain a yield of less than 12%
The present value ratio will be
$$\frac{977\ 598}{1\ 000\ 000} = 0.977598$$

If the developer had been able to obtain a yield of twelve per cent on reinvestment the present value would have been $1 069 762, identical to that in Example 10.

There may be occasions when the cash flow contains negative numbers, say in the early years of a development when it is expected to lose money before it starts earning a profit. In this case the present values of the negative

YIELD ANALYSIS

incomes are deducted from the present values of the positive incomes rather than added to the cost when assessing the total present value.

Example 12: A developer has a development which has a total capital cost of $1 000 000 and which he expects to sell for $1 600 000 after five years. The anticipated net income is as follows:

Year	
1	$ 20 000 loss
2	5 000 loss
3	75 000
4	100 000
5	120 000

Will he obtain a yield of at least 10%?

Answer: The discounted cash flow will be:

Year	Net Cash flow × V @ 10%	=	Present Value
1	− $ 20 000 × 0.909090909	=	− 18 182Cr
2	− $ 5 000 × 0.826446281	=	− 4 132Cr
3	$ 75 000 × 0.751314801	=	56 349
4	$ 100 000 × 0.683013455	=	68 301
5	$ 120 000 × 0.620921323	=	74 511
5 (Sale)	1 600 000 × 0.620921323	=	993 474
Total Present Value			$1 170 321

The total present value exceeds the cost so the developer will receive a yield of at least 10%.

The present value ratio will be:

$$\frac{1\ 170\ 321}{1\ 000\ 000} = 1.170321$$

When a mortgage is needed to help finance the project the calculations are complicated by having to work out the balance remaining on the mortgage, if any, but otherwise they are similar to those already described.

Example 13: Will the developer obtain a yield of at least 12% for the development shown in Table 8 if he expects to sell it after ten years for $1 500 000?

Answer: In ten years the developer will have repaid $106 212 of the mortgage principal, leaving a balance of $643 788 still to be repaid. Assuming he can pay off the mortgage at that time or have the new owner assume it, his income from the sale will be:

Sale price	$1 500 000
Less balance of mortgage to be paid off	643 788
Income to the developer	$ 856 212

```
Income = $13 372 × YP, 10 years @ 12%
       =  $13 372 × 5.650223028           =    $75 555
 Sale     $856 212 × V, 10 years @ 12%
       = $856 212 × 0.321973237           =    275 677
       Total Present Value                     $351 232
```

The total present value is higher than the owner's equity ($250,000) so the development will yield at least 12%.

The present value ratio will be:

$$\frac{351\ 232}{250\ 000} = 1.404928$$

This is a considerably better ratio than was found for Example 8 which was similar but had no mortgage to help with the financing, and shows how leverage helps to improve the yield.

An alternative method sometimes used to determine net present value is to add the present value of all the positive incomes, and deduct the present value of all the negative incomes including the capital cost or investment. As long as the answer is positive the yield will be met, but if the answer is negative the net present value is less than the investment so the development will not give the required yield. When this approach is used a net present value ratio is not calculated.

Internal Rate of Return

Capitalization methods will give the yield under certain conditions, and net present value calculations will show whether a given yield can be obtained, but neither will show what the actual yield is likely to be under any conditions. To do this the internal rate of return must be found.

The internal rate of return is that rate of interest which makes the present value of future income equal to the total investment, so the calculations are very similar to those used in the net present value approach. The principal difference between the two is that whereas with the net present value approach the yield is given, when solving for the internal rate of return the yield has to be found.

There are several formulae which can be used to find the internal rate of return, all of which give the same answer and all of which are based on the fact that a rate must be found which makes the present value of future income equal to the total investment. A variation of this is to find the rate which makes the present value of future income divided by the total investment equal to one and this method, which requires the same calcuation as was used to find the present value ratio, will be used in the following examples.

The internal rate of return cannot be calculated directly, but has to be derived by interpolation between two rates which are believed to straddle the internal rate of return. This means that the present value at two assumed

YIELD ANALYSIS

capitalization (yield) rates have to be calculated in the same way as was done in the net present value approach and the internal rate of return found between them.

Example 14: What is the internal rate of return (yield) for the development shown in Table 7 if the developer expects to sell the development after ten years for $1 500 000?

Answer: The circumstances of this development are the same as those shown in Examples 4 and 8. Example 4 showed that the yield would be at least ten per cent, while Example 8 showed that it will exceed twelve per cent. The two rates which will be tested therefore are twelve per cent, which is known to be less than the yield, and fifteen per cent, which is likely to be more than the yield, and the internal rate of return interpolated between them.

		Present Value @ 12%	Present Value @15%
Income	$100 000 × YP, 10 years @ 12% $100 000 × 5.650223028 =	$565 022	–
	$100 000 × YP, 10 years @ 15% = $100 000 × 5.018768626 =	–	$501 877
Sale	$1 500 000 × V, 10 years @ 12% = $1 500 000 × 0.321973237 =	482 960	–
	$1 500 000 × V, 10 years @ 15% = $1 500 000 × 0.247184706 =	–	370 777
Total Present Value		$1 047 982	$872 654
Present Value Ratio		$\frac{1\,047\,982}{1\,000\,000}$ = 1.047982	$\frac{872\,654}{1\,000\,000}$ = 0.872654

Since the internal rate of return is that rate where the present value ratio equals one, the internal rate of return must be between twelve and fifteen per cent. The actual rate is found by using the following formula:

$$IRR = L\% + \frac{HR - 1.0}{HR - LR}(H\% - L\%)$$

where HR = the higher ratio
LR = the lower ratio
H% = the higher capitalization rate
and L% = the lower capitalization rate.

This becomes:

$$IRR = 12\% + \frac{1.047982 - 1.0}{1.047982 - 0.872654} \times (15-12)$$

CONSTRUCTION ESTIMATING AND COSTING

$$= 12\% + \frac{0.047982}{0.175328} \times 3$$
$$= 12\% + 0.82101$$
$$= 12.82\%$$

The yield rate is (say) $12^3/_4\%$

The formula used in this example gives a straight-line interpolation which is generally considered to be accurate enough in view of the fact that the incomes on which the present values have been calculated are estimated figures only. It is, in any event, highly unlikely that an accuracy closer than one per cent will be achieved in practice.

Example 15: If the developer decides to hold the property for 40 years, at which time he expects it to be demolished, what will the internal rate of return (yield) be?

Answer: This is the same situation as occurred in Example 5 where the balance of income attributable to the building was found to be $90 000 per year when the return required on the land was discounted at 10%, and the yield was found to be less than ten per cent. The rates which will be tested are eight per cent and ten per cent.

	Present Value @ 8%	Present Value @ 10%
Income $92 000 × YP, 40 years @ 8%		
= $92 000 × 11.92461333 =	1 097 064	–
$90 000 × YP, 40 years @ 10%		
= $90 000 × 9.779050719 =	–	880 115
Present value of land	100 000	100 000
Total present value	$1 197 064	$980 115
Present Value Ratio	1 197 064	980 115
	1 000 000	1 000 000
=	1.197064	0.980115

$$\text{IRR} = L\% + \frac{HR - 1.0}{HR - LR}(H\% - L\%)$$

$$= 8\% + \frac{1.197064 - 1.0}{1.197064 - 0.980115} \times (10 - 8)$$

$$= 8\% + \frac{0.197064}{0.216949} \times 2$$

YIELD ANALYSIS

$$= 8\% + 1.816685$$

$$= 9.82\%$$

The yield is (say) $9^{3}/_{4}\%$

Note that when calculating the present value at 8% the return on the land is discounted at 8% leaving a balance of income attributable to the building of $92 000.

Modified Internal Rate of Return

As with the present value approach, solving for the internal rate of return makes the implicit assumption that the income derived from the investment will be reinvested at the same yield rate as the internal rate of return. While the internal rate of return is the true yield, if it is found to be higher than the rate at which the income can be reinvested, or where a comparison is to be made between two or more investments which are spread over different time periods, a modified internal rate of return needs to be calculated.

The modified internal rate of return is found by compounding the income forward at the assumed reinvestment rate to give a future value, and then finding that rate which will discount the future value back to equal the cost of the investment.

Example 16: What is the modified internal rate of return for the development shown in Table 7 if the developer expects to sell the development after ten years for $1 500 000 and his reinvestment rate will be 10%?

Answer: Income $100 000 × FV, 10 years @ 10%

$100 000 × 15.9374246 =	$1 593 742
Sale	1 500 000
Total Future Value	$3 093 742
Capital Cost	$1 000 000

The discount rate which will give a future value of $3 093 742 in ten years' time for a capital cost of $1,000,000 will be:

$$\sqrt[10]{\frac{3\ 093\ 742}{1\ 000\ 000}} - 1$$

$$= 1.11956267 - 1$$

$$= 11.956267\%$$

The modified internal rate of return is 11.96%.

This compares with the internal rate of return of 12.82% in Example 14 where the implicit assumption was made that the income would be reinvested at 12.82%.

A Small Office Building

Table 9 shows a pro-forma for a small office building development which has an irregular cash flow income. Periodic redecorations are expected to occur every five years and periodic repairs every ten years, and the net annual income is reduced for the first twenty-five years by the amount of the mortgage repayments. If the developer were to hold the development for more than twenty-five years his income would increase by $72 767 a year beyond that time.

On the assumption that the developer will wish to sell the development at the end of ten years, the cash flow during the term of his investment will be:

Year			
1	$161 760 − $142 567	=	$19 193
2	$161 760 − $142 567	=	$19 193
3	$161 760 − $142 567	=	$19 193
4	$161 760 − $142 567	=	$19 193
5	$161 760 − $142 567 − $4 000	=	$15 193
6	$161 760 − $142 567	=	$19 193
7	$161 760 − $142 567	=	$19 193
8	$161 760 − $142 567	=	$19 193
9	$161 760 − $142 567	=	$19 193
10	$161 760 − $142 567 − $4 000 − $3 000	=	$12 193

TABLE 9
PRO-FORMA FOR A SMALL OFFICE BUILDING
Capital Cost

1. Land	$ 60 000
2. Demolition	3 000
3. Construction and site works	600 000
4. Land surveys	1 500
5. Soil tests	1 500
6. Market studies	4 000
7. Appraisal fees	—
8. Design fees	48 000
9. Special consultants' fees	—
10. Legal fees	8 000
11. Accounting fees	—
12. Underwriter's fees	—
13. Developer's fee	10 000
14. Project management fee	—
15. Furnishings	—

YIELD ANALYSIS

16. Artwork		—
17. Municipal levies		—
18. Parking levies		—
19. Capital taxes		—
20. Real estate taxes		2 500
21. Insurances		—
22. Mortgage insurance		—
23. Performance bonds		—
24. Maintenance costs		—
25. Accommodation		—
26. Developer's administrative costs		3 000
27. Leasing commissions		16 000
28. Advertising and public relations		2 000
29. Tenant inducements		18 000
30. Tenant lease takeovers		—
31. Interim lender commitment fees		—
32. Long-term lender standby fees		—
33. Brokerage commission		6 500
34. Registration fees		—
35. Interim finance		36 000
36. Initial occupancy costs		—
37. Contingencies		10 000
		$ 830 000
38. Income during construction		—
Total Capital Cost		$ 830 000

Gross Annual Income

Office space	1115 m² @$96.86	108 000
Ground floor space	372 m² @ $129.03	48 000
Parking	12 No. @ $480.00	5 760
		$ 161 760

Annual Expenses

1. Real estate taxes	22 000
2. Insurance	6 000
3. Cleaning and supplies	12 000
4. Hydro	4 500
5. Fuel	2 500
6. Water	1 000
7. Security	—
8. Operating staff	5 000
9. Service contracts	4 700
10. Supplies	1 500
11. Building management	—

12. Garbage removal	—
13. Snow removal	500
14. Gardening	—
15. Land rental	—
16. Preventive maintenance	—
17. Vacancy allowance	8 100
18. Legal and audit	2 000
19. Communication equipment	—
20. Finance costs ($630 000 @ 11% for 25 years)	72 767
	$142 567

Periodic Expenses

1. Redecorations	$4000 every 5 years
2. Repairs	$3000 every 10 years

Owner's Equity

Total capital cost	$830 000
Less Mortgage	630 000
	$200 000

Since the developer expects to sell the development before the mortage has expired, the balance of the mortage principal will have to be calculated as was done in Example 13. On a mortgage of $630 000 at eleven per cent with an amortization period of twenty-five years, the principal repayments at the end of ten years will be $89 217, so the remaining balance will be $540 783. If the expected sale price at the end of ten years is $950 000, the developer's income from the sale, assuming he can discharge the mortgage, will be:

	Sale Price	$950 000
Less	Balance of mortgage	540 783
	Net proceeds of the sale	$409 217

With the incomes shown for this development it is difficult to determine the target rates for calculating the internal rate of return. Using a simple capitalization method with the owner's equity of $200 000 as the capital cost and taking $19 000 as an average annual income, a target rate might be:

$$\frac{\$19\,000}{\$200\,000} \times 100 = 9.50\%$$

This, however, ignores the large income the developer will receive as a result of the sale at the end of ten years. An alternative way of calculating a target rate might be to find the average annual income including the proceeds of the sale and use this figure:

Total annual income	$180 930
Proceeds from the sale	409 217
	$590 147

YIELD ANALYSIS

Average annual income over ten years = $59 015

$$\frac{\$59\ 015}{\$200\ 000} \times 100 = 29.51\%$$

All this shows is that the internal rate of return probably lies somewhere between 9.50% and 29.51%, which isn't very helpful.

Assume target rates of ten and fifteen per cent:

Year	Income	Present Value @ 10%	Present Value @ 15%
1	$19 193	$17 448	$16 690
2	$19 193	$15 862	$14 513
3	$19 193	$14 420	$12 620
4	$19 193	$13 109	$10 974
5	$15 193	$ 9 434	$ 7 554
6	$19 193	$10 834	$ 8 298
7	$19 193	$ 9 849	$ 7 215
8	$19 193	$ 8 954	$ 6 274
9	$19 193	$ 8 140	$ 5 456
10	$12 193	$ 4 701	$ 3 014
10 (Sale)	$409 217	$157 771	$101 152
Total Present Value		$270 522	$193 760
Present Value Ratio		$\frac{270\ 522}{200\ 000}$	$\frac{193\ 760}{200\ 000}$
	=	1.3526	0.9688

The internal rate of return is therefore between ten and fifteen per cent, and nearer to fifteen per cent than to ten per cent. At this point with the help of a computer or a calculator with financial functions the actual rate could be found by repeatedly testing at different rates until the present value ratio equals one. Alternatively a third rate, say fourteen per cent, can be tried and if its present value ratio is found to exceed one, the actual rate can be interpolated between fourteen and fifteen per cent:

Year	Income	Present Value @ 14%
1	$ 19 193	$ 16 836
2	$ 19 193	$ 14 768
3	$ 19 193	$ 12 955
4	$ 19 193	$ 11 364
5	$ 15 193	$ 7 891
6	$ 19 193	$ 8 744

7	$ 19 193	$ 7 670
8	$ 19 193	$ 6 728
9	$ 19 193	$ 5 902
10	$ 12 193	$ 3 289
10 (Sale)	$409 217	$110 384
Total Present Value		$206 531

Present Value Ratio $= \dfrac{206\,531}{200\,000} = 1.032655$

$$\text{IRR} = L\% + \frac{HR - 1.0}{HR - LR}(H\% - L\%)$$

$$= 14\% + \frac{1.032655 - 1.0}{1.032655 - 0.9688} \times (15 - 14)$$

$$= 14\% + \frac{0.032655}{0.063855}$$

$$= 14.511393\%$$

The yield rate is (say) 14½%.

If the developer were to hold the development for, say, forty years, at the end of which time the building might be expected to be demolished, he would have to prepare a cash flow covering forty years based on the example just given. This is just what he would have to do if taxation were included in the calculations, but since taxes are not a factor in the present examples there is, fortunately, a shorter way of finding the internal rate of return over forty years.

This involves separating the net income for the first twenty-five years from the net income for the final fifteen years when mortgage payments are no longer required, and treating the periodic expenses as negative income. The net income for the final fifteen years is discounted back to give a present value in twenty-five years' time, and then this sum is further discounted back to give a present value today. The present value factors for the periodic expenses for each of the years in which they occur are totalled so as to find their present value. The following example will, it is hoped, make this clear.

Assume target rates of ten and twelve per cent:

	Present Value @ 10%	Present Value @ 12%
Income for the first 25 years: $19 193 × YP, 25 years @ 10% = $19 193 × 9.077040018 =	174 216	—

YIELD ANALYSIS

$19 193 × YP, 25 years @ 12%
= $19 193 × 7.843139113 =   150 533
Income for the final 15 years:
    $91 960 × YP, 15 years @ 10%
= $91 960 × 7.606079506 = $699 455
    $699 455 × V, 25 years @ 10%
= $699 455 × 0.092295998 =   64 557   —
    $91 960 × YP, 15 years @ 12%
= $91 960 × 6.810864489 = $626 327
= $626 327 × V, 25 years @ 12%
= $626 327 × 0.058823307 =   —   36 843

Periodic expenses:

Year	10%		12%	
5	0.6209	—	0.5674	—
10	0.3855	0.3855	0.3220	0.3220
15	0.2394	—	0.1827	—
20	0.1486	0.1486	0.1037	0.1037
25	0.0923	—	0.0588	—
30	0.0573	0.0573	0.0334	0.0334
35	0.0356	—	0.0189	—
	1.5796	0.5914	1.2869	0.4591

Periodic redecorations
$4 000 × 1.5796 =   6 318 cr   —
$4 000 × 1.2869 =   —   5 148 cr

Periodic repairs
$3 000 × 0.5914 =   1 774 cr   —
$3 000 × 0.4591 =   —   1 377 cr

Residual land value
    $60 000 × V, 40 years @ 10%
= $60 000 × 0.022094928   =   1 326   —
    $60 000 × V, 40 years @ 12%
= $60 000 × 0.010746798   =   —   645

    Total Present Value $232 007 $181 496

    Present Value Ratio $\dfrac{232\ 007}{200\ 000}$ $\dfrac{181\ 496}{200\ 000}$

                                        =   1.16004   0.90748

$$\text{IRR} = L\% + \frac{HR - 1.0}{HR - LR}(H\% - L\%)$$

$$= 10\% + \frac{1.16004 - 1.0}{1.16004 - 0.90748} \times (12 - 10)$$

$$= 10\% + \frac{0.16004}{0.25256} \times 2$$

$$= 10\% + 1.26734$$
$$= 11.26734\%$$

The yield rate is (say) $11^{1}/_{4}\%$.

Escalation

Mention was made earlier in this chapter to the fact that a number of assumptions have to be made when calculating the yield on an investment property. One of the assumptions which could be made is the probable inflation of income due to increased rental rates in the future. However, coupled with increased rentals is likely to be increases in operating costs. It is usual therefore to base the yield analysis on the known, or reasonably assumed facts at a particular point in time, usually the time when construction is completed and the building is available for rent, and to ignore any probable future escalation.

CHAPTER 11

Taxation

Although taxation has not been included in any of the yield calculations shown in the last chapter, this is not to say that it is not of the greatest concern to the investor. The inclusion of the tax man as a partner in an investment gives an added dimension to the calculations, invariably reducing the yield, and possibly in some instances making a marginally profitable investment into a decidedly unprofitable one.

The way in which taxes are to be calculated and the actual rate of tax are subject to change from time to time, but the manner of incorporating them into a present value calculation will remain basically the same regardless of any revisions to tax legislation made by governments from time to time. The examples given later in this chapter should therefore be looked upon as showing the method of incorporating taxes into the calculations, but not necessarily reflecting the current tax rates which should be used.

Corporation Tax

Anyone who initiates a building development will normally do so through a corporation rather than as an individual because this has certain tax advantages. Either an existing development company will be used or, as frequently happens, a new company will be incorporated expressly for the development.

Under the Federal Income Tax Act the federal government levies taxes on all Canadian residents and non-residents, whether corporations or individuals, carrying on business in Canada. In addition, each of the provinces has its own act which imposes taxes on its residents, and on non-residents who have business establishments there.

The current federal tax rate on a corporation is forty-six per cent, which is reduced by ten per cent for corporations paying provincial taxes. The provincial tax on a corporation varies from ten to thirteen per cent, depending upon the province, making the effective annual tax rate somewhere between forty-six and forty-nine per cent.

In most instances taxpayers file tax returns with the federal government for both federal and provincial taxes. The federal government then collects both taxes and passes on the provinces' portion to them. The exceptions to this are corporations within Ontario and Quebec which must file separate provincial tax returns.

Small Business Deduction

Section 125 (1) of the Income Tax Act permits a Canadian-controlled private corporation to deduct, under certain conditions, a percentage, currently twenty-one per cent, of business income from tax otherwise payable. The effect of this is to reduce the effective tax rate from forty-six to twenty-five per cent of taxable income, which is better than many people's personal tax rate.

A Canadian-controlled private corporation is defined by the act as one which is not a public corporation, which is not controlled directly or indirectly by one or more public corporations, nor by one or more non-resident corporations, and which is resident in Canada.

The small business deduction is allowed only on income from an active business carried on in Canada and not on investment income. Although the words "active business" are not defined in the act a development company which maintains a management office and full-time management staff while deriving income from the rental of a building is deemed to be carrying on an active business and can, therefore, take advantage of the small business deduction.

The rules governing the small business deduction are subject to change periodically, but currently are basically as follows:

1. The maximum amount on which the deduction may be calculated is $150 000 in any one year. This means that if the taxable income during the year is $150 000 or less the effective tax rate is twenty-five per cent. However, if it is more than $150 000 the deduction can only be taken on $150 000, the balance being taxed at the full rate of forty-six per cent.
2. The total taxable income earned since 1971 together with four-thirds of the dividends received from other taxable Canadian corporations are accumulated into what is known as the cumulative deduction account. This account is used as a further control on the small business deduction because the deduction cannot be taken on any income which makes the cumulative deduction account exceed $750 000. The cumulative deduction account is made up of the total taxable income earned over the years, not just the income which is subject to the small business deduction. This means that every dollar of taxable income in excess of $150 000 each year not only is not eligible for the small business deduction in that year but

TAXATION

also will increase the cumulative deduction account and hence reduce the small business deduction in future years.
3. The cumulative deduction account can be reduced by an amount equal to four-thirds of the dividends paid out by the corporation since 1971. The reduction of the cumulative deduction account has the effect of extending the period of time over which the small business deduction can be claimed. The cumulative deduction account can also be reduced by a sum equal to four times the amount by which the corporation's refundable dividend tax on hand for the year exceeds its dividend refund, but this situation is unlikely to arise in a development company.

Capital Cost Allowances

The capital cost of a new building is not deductible as an expense for tax purposes, but as with many other assets depreciation is allowed in the form of a capital cost allowance over the life of the asset. The Federal Income Tax Act has established various classes of depreciable assets, each of which can be depreciated at a differing rate annually by means of a capital cost allowance. Real estate is one class, but every rental property with a capital cost of more than $50 000 must be kept separate so that the capital cost allowance can be applied to each individual building. This is to facilitate the adjustment of the capital cost allowance in the event of a sale.

The capital cost allowance is claimed on the diminishing balance method which means that the depreciation rate is applied each year to the original cost of the building less the accumulated total of capital cost allowances claimed in previous years. While this is the method used for tax purposes there are other ways in which a company can allow for depreciation, so the capital cost allowance is not necessarily the same as the amount of depreciation carried in the company's books. The difference between the original capital cost and the total capital cost allowances claimed is known as the unclaimed capital cost.

If the property is subsequently sold for a price which is more than the unclaimed capital cost, then the difference between the two is added in to taxable income for that year. However, this is subject to a maximum which is equal to the total amount of capital cost allowance claimed. In addition, if the proceeds of the sale exceed the original capital cost this is a capital gain, half of which must be added to taxable income.

If the property is subsequently sold for a price which is less than the unclaimed capital cost, then the difference between the two is deducted from taxable income for that year as a terminal loss and there would, of course, be no capital gain.

It should be noted that the maximum capital cost allowance in any year cannot exceed the rental income from a rental building unless the owner is a corporation whose principal business is holding, developing or selling real estate. This is to prevent taxpayers, particularly individuals in high tax brackets, from investing in depreciable real estate and using the capital cost allowance as a means of reducing tax on other income.

Internal Rate of Return

With a knowledge of the basis of taxation, and the tax rate to be used, it only remains for cash flows incorporating taxes to be set up so the internal rate of return can be calculated. Although the cash flow is required as a preliminary to calculating the internal rate of return, the same cash flow would be required if the net present value approach were to be used.

Example 1: What is the internal rate of return for the development shown in Table 7 if the developer expects to sell the development after ten years for $1 500 000, assuming that the development will be done through a corporation and that taxes have to be included in the calculations?

Answer: This is similar to Example 14 in the last chapter except that taxes now have to be included. With a net annual income of $100 000 the development comes within the orbit of the small business deduction which means that the effective federal tax rate is 25%. However, this can be reduced by ten per cent for provincial taxes, and if it is assumed that there are no provincial tax abatements and the effective provincial tax rate is 12%, the total effective tax rate will be 27%.

In Example 14 the total present value could be found with comparative ease, but now that taxes have to be included a cash flow schedule has to be set up as follows:

Year	Net Income $	Capital Cost Allowance $	Taxable Income $	Tax Payable $	After Tax Cash Flow $
1	100 000	45 000	55 000	14 850	85 150
2	100 000	42 750	57 250	15 458	84 542
3	100 000	40 613	59 387	16 034	83 966
4	100 000	38 582	61 418	16 583	83 417
5	100 000	36 653	63 347	17 104	82 896
6	100 000	34 820	65 180	17 599	82 401
7	100 000	33 079	66 921	18 069	81 931
8	100 000	31 425	68 575	18 515	81 485
9	100 000	29 854	70 146	18 939	81 061
10	100 000	28 361	71 639	19 343	80 657
		361 137	638 863		
10 (Sale)	1 500 000	361 137 cr	361 137 250 000 }	276 889	1 223 111

The depreciation rate taken for the capital cost allowance is five per cent of the capital cost of the building only, on the assumption that it will be of masonry or concrete. If it were wood or metal the depreciation rate would be ten per cent. Because the sale price in Year Ten is more than the unclaimed capital cost the total of the capital cost allowances has to be included as

TAXATION

taxable income in that year. In addition, half the difference between the original capital cost of $1 000 000 and the sale price of $1 500 000 also has to be included as taxable income.

It has been assumed that the corporation will pay no dividends during the ten years so the cumulative deduction account will be the same as the taxable income, that is $638 863 up to the time of the sale. When the sale has been completed another $611 137 ($361 137 plus $250 000) is added onto taxable income, and this will increase the cumulative deduction account to a point where it exceeds the maximum of $750 000. However, the sale also increases the income for year ten above the allowable $150 000. This is, in fact, the critical calculation so the tax payable on the sale will be:

$150 000 (maximum allowable) − $71 639 (taxable income in year 10) = $78 361 (maximum income at lower rate)

$78 361 @ 27% =	$21 157
$611 137 − $78 361 = $532 776	
$532 776 @ 48%	$255 732
Total	$276 889

In practice the annual income and the proceeds of the sale would not be shown as separate amounts in Year Ten but would be consolidated into one, so that net figures would be given for the capital cost allowance, taxable income, tax payable and after-tax cash flow.

Having established the after-tax cash flow the internal rate of return can be calculated using the procedure described in the last chapter. Example 14 gave an internal rate of return of 12.82%, but now that tax has been incorporated the rate is bound to be somewhat lower. The two rates to be tested will therefore be eight per cent and ten per cent.

Year	Income	Present Value @ 8%	Present Value @ 10%
1	85 150	78 843	77 409
2	84 542	72 481	69 869
3	83 966	66 655	63 085
4	83 417	61 314	56 975
5	82 896	56 418	51 472
6	82 401	51 927	46 513
7	81 931	47 806	42 044
8	81 485	44 024	38 013
9	81 061	40 551	34 378
10	1 303 768	603 897	502 659
Total Present Value		$1 123 916	$982 417
Present Value Ratio		$\frac{1\,123\,916}{1\,000\,000}$	$\frac{982\,417}{1\,000\,000}$
	=	1.123916	0.982417

$$IRR = L\% + \frac{HR - 1.0}{HR - LR}(H\% - L\%)$$

$$= 8\% + \frac{1.123916 - 1.0}{1.123916 - 0.982417} \times (10 - 8)$$

$$= 8\% + \frac{0.123916}{0.141499} \times 2$$

$$= 8\% + 1.751475$$

$$= 9.75\%$$

The yield rate is $9^3/_4\%$.

Comparing this with the yield found for Example 14 in the last chapter shows that the effect of taxation has been to reduce the yield by almost exactly three per cent.

When a mortgage is used to help finance the project the calculations are essentially the same, but become a little more complicated.

Example 2: What is the internal rate of return for the development shown in Table 8 if the developer expects to sell the development after ten years for $1 500 000, assuming that the development will be done through a corporation and that taxes have to be included in the calculations?

Answer: This will be tackled in the same way as Example 1, assuming the same tax rate and starting with a cash flow schedule, although the schedule will now have to incorporate the mortgage payments in it as follows:

Year	Net Income $	Mortgage Interest $	Mortgage Principal $	Mortgage Balance $	Capital Cost Allowance $	Taxable Income $	Tax Payable $	After Tax Cash Flow $
1	13 372	80 368	6 260	743 740	19 632	—	—	13 372
2	13 372	79 660	6 967	736 773	20 339	—	—	13 372
3	13 372	78 873	7 755	729 018	21 127	—	—	13 372
4	13 372	77 996	8 631	720 387	22 003	—	—	13 372
5	13 372	77 021	9 606	710 781	22 978	—	—	13 372
6	13 372	75 935	10 693	700 088	24 065	—	—	13 372
7	13 372	74 727	11 901	688 187	25 273	—	—	13 372
8	13 372	73 382	13 246	674 941	26 618	—	—	13 372
9	13 372	71 884	14 743	660 198	28 115	—	—	13 372
10	13 372	70 218	16 410	643 788	29 782	—	—	13 372
					239 932			
10 (Sale)	1 500 000	—	—	643 788	239 788 cr	239 932 } 250 000	203 667	652 545

If the net income had been sufficiently high during the ten-year holding period the taxable income would have been calculated by adding the mortgage principal to the net income, and then deducting the capital cost allowance as was done in the last example. The mortgage principal has to be added to the net income because while the mortgage interest can be claimed as a tax deduction, the principal repayment cannot, so it has to be added back as part of the income.

In this example deducting a five per cent capital cost allowance from the net income plus the mortgage principal results in a negative taxable income. In some circumstances it might be worthwhile using the negative taxable income (that is, the tax losses) to offset profits elsewhere, or carrying it forward into future years, but in this instance it seems preferable to reduce the capital cost allowance each year down to a figure which makes the taxable income zero. The capital cost allowance is then not five per cent of the depreciated capital cost, but the sum of the net income plus the mortgage principal repayment. This also has the effect of reducing the total capital cost allowance so that when the building is sold the taxable income is less than it would have been if the full capital cost allowance had been claimed.

The calculation of the taxable income where the property is sold is done in the same way as in Example 1, but the calculation of the tax payable is as follows:

Total taxable income, $239 932 + $250 000	=	$489 932
Tax on the first $150 000 @ 27%	=	$40 500
Tax on the balance, $339 932 @ 48%	=	163 167
		$203 667

From this, the after tax cash flow will be:

Proceeds of the sale		$1 500 000
Less Mortgage balance		643 788
		$856 212
Less Tax payable		203 667
		$652 545

Because the income is constant for ten years, unlike the income in Example 1, the calculation of the internal rate of return is comparatively simple. Using target rates of twelve and fourteen per cent:

	Present Value @ 12%	Present Value @ 14%
Income $13 372 p.a. for 10 years	75 555	69 750
Sale $652 545 in 10 years	210 102	176 020
Total Present Value	$285 657	$245 770

CONSTRUCTION ESTIMATING AND COSTING

	Present Value @ 12%	Present Value @ 14%
Present Value Ratio	$\dfrac{285\ 657}{250\ 000}$	$\dfrac{245\ 770}{250\ 000}$
=	1.142628	0.98308

$$\text{IRR} = L\% + \frac{HR - 1.0}{HR - LR}(H\% - L\%)$$

$$= 12\% + \frac{1.142628 - 1.0}{1.142528 - 0.98308} \times (14 - 12)$$

$$= 12\% + \frac{0.142628}{0.159548} \times 2$$

$$= 12\% + 1.7879$$

$$= 13.7879$$

The yield rate is (say) $13^{3}/_{4}\%$.

CHAPTER 12
Life-Cycle Costing

One of the objectives given in chapter five for cost control in the pre-tender period of a construction program was that the building owner should receive value for his money. However, the reference there was to capital cost, and while capital cost is important, it is only one of the costs associated with a building over its life. Costs continue throughout the life of a building in the form of maintenance, operating and other running costs, and buildings generally have a long life over which these costs can be incurred. Value for money should therefore extend beyond capital cost to include these other costs. To make decisions on the basis of capital costs alone, without considering these subsequent costs, can be misleading.

Life-cycle costing, also known as life costing, life-span costing, costs-in-use or ultimate costs, is a technique which takes into account the total costs, both present and future. It is a comparative technique, used to help make a decision between choices and on its own a life-cycle cost is relatively meaningless. It can be used, for example, to help decide whether to renovate an existing building or to demolish it and build a new one, or whether a better choice would be to sell the property and rent accommodation elsewhere. In the early design stages it can be used to help select the most appropriate size and shape for the building or the best interior planning solution, by comparing not only the capital costs but also the effect on future operating costs. In the later design stages it can be used to make decisions about the choice of systems and materials in the building. Life-cycle costing can therefore be used to help make decisions throughout the pre-construction period — from decisions about the building as a whole down to decisions about the choice of materials to be incorporated within it.

When life-cycle cost studies are made the most economical solution is usually that which has the lowest life-cycle cost, but there may be occasions when the difference is marginal or when the capital expenditure required for the optimum solution is more than the building owner is prepared to pay. In these circumstances the decision is likely to be to accept the solution which has the lower capital cost, even though it may show a higher life-cycle cost, but at least the owner makes his decision knowing the probable cost implications over the life of the building. However, life-cycle cost studies will frequently show that the most economical design is not necessarily that which is the cheapest in capital cost, and it is generally worth paying more in capital cost if the gain in value is more than the extra cost.

It has been said that value for money spent on a building depends upon the three factors of appearance, function and cost, all three of which are based on judgement. The first is, perhaps, a subjective assessment but nevertheless can be agreed upon between the building owner and the architect. The second is a judgement which must be based upon the owner's requirements, while the third is a judgement which must take into account the other two and strike a proper balance between them and the owner's financial limitations. Life-cycle costing enables this to be done over the life of the building.

Balancing cost against appearance will depend upon the owner's objectives. Some owners are prepared to spend more, both in capital and in running costs, on appearance if it enhances the building's prestige. This may be due to the increased revenue the building can then command, to the greater ease in raising capital for it, to the advertising advantages it gives, or for no good reason except the owner's own sense of importance. Estimating the monetary value of these benefits is often difficult. Balancing cost against function will also depend upon the building owner's objectives. In the case of a manufacturing plant, the cost of the building can be looked on as being one of the factors of production, with no value in itself except as it contributes to the value of the business. In this situation the objective will be to maximize the profits of the business as a whole, which may not necessarily mean minimizing the capital cost of the building. The life-cycle cost of the building in such a case would then include, as it would for a health or education facility, all the costs associated with the functions being carried on within the building as well as the costs involved in operating the building itself.

The value of a building erected for rental purposes, on the other hand, is related solely to its income potential as a building; it is not just one of the factors of production but the only factor of production, and it is unlikely that a developer would concern himself about the functions being performed within it except insofar as they might affect its income potential. In this case the life-cycle costs of the building would not normally need to include the costs of the functions within the building, but only those costs related to the operation of the building.

Present Value Method

Life-cycle costing requires the accumulation of all the costs associated with whatever aspect of the building is being studied, both present and future, to give a total cost. Since it is a comparative method those costs which are common to both sides do not need to be included in the study. A comparison between the costs of two types of floor finishes, for example, would not need to include the cost of a topping if the same topping were required for both. Also, life-cycle costing is concerned with the comparison of costs and rarely needs to take income into account. Only if the choice is likely to affect the income of the building does income need to be included in the calculations.

As was made clear in earlier chapters, adding present and future costs together without discounting the future costs will give a misleading answer because expenditures to be made in the future are not the same as current expenditures. The PRESENT VALUE METHOD of calculating life-cycle costs is therefore very similar to the method described in chapter ten for calculating net present value, in which a discount rate is assumed to find the present value of future expenditures. The discounted values of future expenditures, both annual and periodic, are then added to the capital cost to give a total life-cycle cost. The PRESENT VALUE METHOD is not the only way in which life-cycle costs can be analyzed. The ANNUAL EQUIVALENT METHOD, which will be described later, can also be used but the PRESENT VALUE METHOD is more usual, particularly in North America. This is because it is a comparatively easy technique and in most circumstances provides an adequate answer.

capital cost of $2 000, a life of twenty years, and an estimated repair cost of $200 every ten years. Roof B has a capital cost of $ 3 500, a life of forty years, and an estimated $200 repair cost every twenty years. The building is expected to last for forty years, and the discount rate is assumed to be ten per cent. Which is the more economical roof finish?

Answer: Roof A is less expensive than Roof B in capital cost. However, Roof A has to be replaced during the life of the building, and the cost and timing of repairs is given. Assuming that inflation does not need to be included in the calculations, the total costs without discounting would be as follows:

	Roof A	Roof B
Initial Cost	$2 000	$3 500
Repair at the end of 10 years	200	—
Replacement at the end of 20 years	2 000	—
Repair at the end of 20 years	—	200
Repair at the end of 30 years	200	—
Total costs	$4 400	$3 700

This would seem to indicate that Roof B is the less expensive over the life of the building, largely because of the replacement cost of Roof A in twenty years' time. If the future expenditures are now discounted to present values the comparison would appear as follows:

CONSTRUCTION ESTIMATING AND COSTING

	Roof A	Roof B
Initial Cost	$2 000	$3 500
Repair at the end of 10 years: $200 × V = $200 × 0.38554 =	77	—
Replacement at the end of 20 years: $2000 × V = $2 000 × 0.14864	297	—
Repair at the end of 20 years: $200 × V = $200 × 0.14864	—	30
Repair at the end of 30 years: $200 × V = $200 × 0.05731 =	11	—
Total present cost	$2 385	$3 530

    The indication now is that Roof A not only has a lower capital cost than Roof B but also a lower life-cycle cost and it is therefore the more economical. Discounting the cost of replacing Roof A in twenty years has reduced its cost from $2 000 to $297 which is the principal reason why Roof A has now become less expense than Roof B.

In this example, which is extremely simple, such variables as the capital costs, the cost of repairs, the discount rate, the life of each roof and the life of the building are given. In practice they are rarely given, they have to be assumed and the reasoning behind the assumptions will be dealt with later. The example also raises a question which has sometimes been asked: Couldn't the building owner invest the difference between the capital costs of the two roofs, $1 500, to earn interest, and shouldn't this be taken into account in the calculation? The answer is that this is a present value calculation and the difference in capital costs is taken into account when the total present costs are compared, so there is no need to make any further adjustment.

Example 2: The cost of two heating systems are to be compared. System A has an intitial cost of $50 000 and an annual operating and maintenance cost of $4 500. System B has an initial cost of $60 000 and an annual operating and maintenance cost of $3 000. Both systems are expected to last for thirty years and the discount rate is assumed to be ten per cent. Which is the more economical system?

Answer: The following is a comparison between the two systems:

	System A	System B
Initial cost	$ 50 000	$ 60 000
Operating and maintenance costs: System A $4 500 × YP, 30 years at 10% = $4 500 × 9.42691 =	42 421	

LIFE-CYCLE COSTING

	System A	System B
System B $3 000 × YP, 30 years at 10% = $3 000 × 9.42691 =		28 281
Total present cost	$ 92 421	$ 88 281

Despite its higher initial cost System B has a lower life-cycle cost over thirty years.

Example 3: The cost of two floor finishes are to be compared. Floor A has an initial cost of $10 000, needs replacing every ten years, and costs $1 200 in maintenance every year. Floor B has an initial cost of $15 000, needs replacing every fifteen years, and has a maintenance cost of $800 a year. The building is expected to last forty years, and the discount rate is assumed to be eight per cent. Which is the more economical floor finish?

Answer:

	Floor A	Floor B
Initial cost	$ 10 000	$ 15 000
Replacement at the end of 10 years: $10 000 × V = $10 000 × 0.46319 =	4 632	
Replacement at the end of 15 years: $15 000 × V = $15 000 × 0.31524 =		4 729
Replacement at the end of 20 years: $10 000 × V = $10 000 × 0.21455 =	2 145	—
Replacement at the end of 30 years (A) $10 000 × V = $10 000 × 0.09938 =	994	—
(B) $15 000 × V = $15 000 × 0.09938 =	—	1 491
Maintenance: Floor A: $1 200 × YP, 40 years @ 8% = $1 200 × 11.92461 =	14 310	—
Floor B: $800 × YP, 40 years @ 8% = $800 × 11.92461 =	—	9 540
Total present cost	$32 081	$30 760

The difference is marginal, the life-cycle cost of Floor B proving to be slightly less than Floor A despite its higher capital cost. The difference is largely accounted for by the difference in maintenance costs, and in this instance the building owner would probably be wiser to select Floor A because of the fifty per cent premium in the initial cost of Floor B, and also in the hope that the maintenance costs of Floor A might in fact be reduced.

An alternative way of making this calculation, which is slightly shorter, is as follows:

	Floor A	Floor B
Initial cost	$10 000	$15 000

Replacement of Floor A:
V @ 10 years = 0.46319
     20 years = 0.21455
     30 years = <u>0.09938</u>
             <u>0.77712</u>

$10 000 × 0.77712 =	7 771	—

Replacement of Floor B:
V @ 15 years = 0.31524
     30 years = <u>0.09938</u>
             <u>0.41462</u>

$15 000 × 0.41462 =	—	6 219

Maintenance:
Floor A: $1 200 × YP, 40 years @ 8%

= $1 200 × 11.92461 =	14 310	—

Floor B: $ 800 × YP, 40 years @ 8%

= $ 800 × 11.92461 =	—	9 540
Total present cost	<u>$32 081</u>	<u>$30 759</u>

It should be noted that in the replacement cycle in Example 3 Floor A is due to be replaced again in year forty. However, since the building has been given a life of only forty years there is no point in replacing the floor in its last year of life. Similarly in Example 1 it was pointless to repair or replace the roofs in the year in which the building was to be demolished.

Present value is a rather imprecise term when used in conjunction with the life-cycle cost of a complete building project. It can mean that point in time when the land is purchased, or the time when construction has been completed and the building is available for occupation – or any point of time in between. Usually it is defined as the time when construction has been completed, in which case all expenditures made before completion should really be discounted forward to give a terminal value at the date of completion so that they can be added to the present value of future expenditures. To do this land and its associated costs would be discounted forward in their entirety, while the cost of construction would be discounted forward from the mid-point of construction to the date of completion. Example 4 illustrates this. However, if the project is small, with a comparatively short time period

LIFE-CYCLE COSTING

between the purchases of the land and completion of the building, this would be an unnecessary added calculation unless the purpose of the life-cycle cost study is to investigate the effect of the construction time on two alternative design proposals.

Example 4: A site has been purchased for $175 000. One alternative is to erect a building which will cost $5 250 000 to build, which will take 12 months to design and 24 months to build, which will have annual expenses of $640 000 and periodic expenses of $75 000 every 25 years. The other alternative is to erect a building which will cost $5 400 000 to build, which will take 10 months to design and 20 months to build, which will have annual expenses of $630 000 and periodic expenses of $50 000 every 20 years. Which alternative should be chosen, assuming a 40 year life in both cases and a discount rate of 9%?

Answer: Although the first alternative is a building with a lower construction cost it must have some peculiarities which cause it to take longer to design and build than the other. The differences in expenses, particularly the periodic expenses, also make it difficult to make a direct comparison between the two. Life-cycle cost studies are made for each alternative in turn:

Alternative 1
The site will be acquired 3 years before construction is complete. The present value of the site, assuming the completion of construction to be the datum and using the future value factor, will therefore be:
$175 000 × A, 3 years @ 9%
= $175 000 × 1.29503   = $ 226 630
The mid-point of construction will be 12 months before construction is complete. The present value of the building, again using the future value factor, will be:
$5 250 000 × A, 1 year @ 9%
= $5 250 000 × 1.09000   = $5 722 500
Design Fees   420 000

Periodic Expenses, 25 years:
 $75 000 × V = $75 000 × 0.11597   8 698

Annual expenses, 40 years:
 $640 000 × YP = $640 000 × 10.75736   = 6 884 710

Total present cost $ 13 262 538

Alternative 2
The site will be acquired 2½ years before construction is complete. The present value of the site will have to be calculated by interpolation as follows:
$175 000 × A
A for 2 years = 1.18810
A for 3 years = <u>1.29503</u>
            2.48313
A for 2½ years = 1.24156
$175 000 × 1.24156 = $ 217 273
The mid-point of construction will be 10 months before construction is complete. The present value of the building will therefore be:
$5 400 00 × A
Interest for one year is 9% so the interest for ⁵/₆ year = .09 × ⁵/₆ = 0.075
A for 10 months = 1.075
$5 400 000 × 1.075 = $5 805 000
Design fees   432 000
Periodic expenses, 20 years:
    $50 000 × V = $50 000 × 0.17843 =   8 922
Annual expenses, 40 years:
    $630 000 × YP = $630 000 × 10.75736 =   <u>6 777 137</u>
            Total present cost   <u>$13 240 332</u>

Despite an additional capital expenditure of $150,000 for Alternative 2, it is in fact less expensive over a 40 year period than Alternative 1. Part of the reason for this is in the difference between the annual expenses for the two alternatives, but by reducing the length of time required for design and construction, Alternative 2 cuts the difference in capital cost from $150 000 to a little over $73 000.

The method of calculating the present value of the site for Alternative 2 by interpolation between the future values for two and three years to give the future value for two and a half years has to be done when discount tables are the only means available for obtaining FUTURE WORTH OF ONE DOLLAR factors. The result is not completely accurate because there is not a straight line relationship between the factors for two adjacent years, although the resulting error is not sufficient to nullify the result. Using a calculator with financial functions the actual factor for the FUTURE WORTH OF ONE DOLLAR

LIFE-CYCLE COSTING

at nine per cent for two and a half years is found to be 1.24041. This would give a present value for the site of $217 072. Similarly, the factor for the future value of the building over a ten month period should in fact be 1.07446 giving a present value of $5 802 064. Making these corrections to Alternative 2 would have the effect of increasing the difference between the two alternatives.

Annual Equivalent Method

Another way of analyzing life-cycle costs is to convert all costs into annual equivalents, using the ANNUAL EQUIVALENT WORTH factor described in chapter nine. Compared to the PRESENT VALUE METHOD this method has the disadvantage that future periodic expenditures must first be converted to present values before they can be expressed as annual equivalent costs, thus adding to the number of calculations which have to be performed. However, there are occasions when it might give a more meaningful answer, as for example when the cost of owning a building is to be compared with the cost of renting equivalent space elsewhere, since rentals are always expressed as annual expenditures. It might also be used when there are no future periodic expenditures to be considered. However, both the PRESENT VALUE METHOD and the ANNUAL EQUIVALENT METHOD will give the same result in terms of ranking the life-cycle costs of different alternatives.

Example 5: Compare the life-cycle costs of the two roofs in Example 1 using the ANNUAL EQUIVALENT METHOD.

Answer: The initial costs can be converted to annual equivalent costs using ANNUAL EQUIVALENT WORTH factors, but before this can be done for the repair and replacement costs they must first be expressed at present values. The calculations will therefore be:

	Roof A	Roof B
Initial cost: Roof A $2 000 × AEW		
= $2 000 × 0.10226 =	$204.52	—
Roof B $3 500 × AEW		
= $3 500 × 0.10226 =	—	$357.91
Repair at the end of 10 years:		
$200 × V = $200 × 0.38554 = $77.11		
$77.11 × AEW = $77.11 × 0.10226 =	7.89	—
Replacement at the end of 20 years:		
$2 000 × V = $2 000 × 0.14864 = $297.28		
$297.28 × AEW = $297.28 × 0.10226 =	30.40	—
Repair at the end of 20 years:		
$200 × V = $200 × 0.14864 = $29.73		
$29.73 × AEW = $29.73 × 0.10226 =	—	3.04

CONSTRUCTION ESTIMATING AND COSTING

Repair at the end of 30 years:
$200 × V = $200 × 0.05731 = $11.46
$11.46 × AEW = $11.46 × 0.10226 =   1.17   —
Total annual equivalent cost $243.98 $360.95

This is rather a lengthy calculation compared to the present value solution found in Example 1. Moreover it can be seen that in each case the present value is first found and then converted to an annual equivalent worth. A far quicker method would be to find the total present cost of each roof as was done in Example 1 and then convert the totals into annual equivalent costs using the ANNUAL EQUIVALENT WORTH FACTOR for 40 years at 10%.

Roof A: Total present cost (from Example 1) = $2 385
Annual equivalent cost = $2 385 × 0.10226 = $243.89

Roof B: Total present cost (from Example 1) = $3 530
Annual equivalent cost = $3 530 × 0.10226 = $360.98

The slight discrepancy in the figures is due to the rounding of the figures in Example 1.

Example 6: Compare the life-cycle costs of the two heating systems in Example 2 using the ANNUAL EQUIVALENT METHOD.
Answer: This example has no future periodic costs so the calculations will be much simpler than in the previous example:

                                                    System A   System B
Initial Cost:
  System A: $50 000 × AEW
          = $50 000 × 0.10608 = $5 304.00   —
  System B: $60 000 × AEW
          = $60 000 × 0.10608 =   — $6 364.80
Annual Costs   4 500.00   3 000.00
Total annual equivalent cost $9 804.00 $9 364.80

Example 7: Compare the life-cycle costs of the two floor finishes in Example 3 using the ANNUAL EQUIVALENT METHOD.
Answer: In this example capital, periodic and annual costs are given. This could require a lengthy calculation but can be shortened slightly as was done in Example 3 in the following manner:

                                                    Floor A   Floor B
Initial cost: Floor A $10 000 × AEW
            = $10 000 × 0.08386 $ 838.60   —
            Floor B $15 000 × AEW
            = $15 000 × 0.08386   — $1 257.90

LIFE-CYCLE COSTING

Replacement of Floor A
V @ 10 years = 0.46319
  20 years = 0.21455
  30 years = <u>0.09938</u>
      0.77712

$10 000 × 0.77712 = $7 771.20
$7 771.20 × 0.08386 = $ 651.69   —
Replacement of Floor B:
V @ 15 years = 0.31524
  30 years = <u>0.09938</u>
      0.41462

$15 000 × 0.41462 = $6 219.30
$6 219.30 × 0.08386 =   —   521.55
Maintenance cost   1 200.00   800.00

Total Annual Equivalent Cost $2 690.29 $2 579.45

Example 8: A small business has a choice between renting space in an existing office building or building its own office building. If it builds, the following costs are expected:
     Cost of the site $60 000
     Other capital costs $770 000
     Annual costs $142 600
     Periodic redecoration $1 500 every 5 years
     Periodic repairs $3 000 every 10 years

If it rents, the annual rental will be $90.00 per m² for 1 200 m², $120.00 per m² for 400 m², and 12 parking spaces at $480.00 each. The rental rates include all taxes etc. If the life of the building is assumed to be thirty years and the discount rate is nine per cent which is the better alternative?

Answer:
Alternative 1 – Construct a New Building
1. Land $60 000 × i = $60 000 × 0.09 = $5 400
2. Other capital costs $770 000 × AEW
     = $770 000 × 0.09734 =   74 952
3. Annual costs   142 600
4. Periodic redecorations
   V @ 5 years = 0.64993
     10 years = 0.42241
     15 years = 0.27454
     20 years = 0.17843
     25 years = <u>0.11597</u>
          1.64128

$1\ 500 \times 1.64128 = \$2\ 461.92$
$2\ 461.92 \times \text{AEW} = \$2\ 461.92 \times 0.09734 =$ 240

5. Periodic repairs:

    V @ 10 years = 0.42241
           20 years = 0.17843
                        0.60084

$3\ 000 \times 0.60084 = \$1\ 802.52$
$1\ 802.52 \times \text{AEW} = \$1\ 802.52 \times 0.09734 =$ 175

Total annual equivalent cost $223 367

Alternative 2 – Rent

1 200 m² @ $90.00 =	$108 000
400 m² @ $120.00 =	48 000
12 Parking spaces @ $480.00 =	5 760
Total annual cost	$161 760

It would therefore appear to be more economical to rent space than to construct a new building. This would certainly be true if inflation does not have to be taken into account. It could be argued that it is highly unlikely that the rental rates will remain constant over the next thirty years, in which case inflation would probably make it more economical to construct a new building. However, the construction alternative carries major annual costs which are also subject to inflation, and the two are likely to offset each other. This question of inflation will be discussed more fully later.

Note, too, that the annual equivalent cost of the land is taken in perpetuity since it is a non-wasting asset and does not need a Sinking Fund allowance.

PRESENT VALUE and ANNUAL EQUIVALENT are not the only ways in which life-cycle costs can be analyzed, although they are the methods most usually used. There is no reason why they shouldn't be analyzed, on FUTURE or TERMINAL VALUES except that this has no advantage over the PRESENT VALUE METHOD and is, perhaps, less meaningful. Another method which might be appropriate under certain circumstances is to select one of the alternatives and to find its total present cost using an assumed discount rate. The other alternative is then analyzed to see what discount rate is needed to give it the same total present cost. This is similar to finding the internal rate of return described in Chapter 10, whichever alternative having the higher discount rate being the more economical. This method will give the same ranking as the PRESENT VALUE and the ANNUAL EQUIVALENT methods, but for a complicated life-cycle cost study it is likely that a computer will be needed for the calculations.

Discount Rates

Because the PRESENT VALUE and ANNUAL EQUIVALENT methods of life-cycle costing require present or future costs to be discounted, the problem of selecting the appropriate discount rate has to be solved. The rate selected will obviously have an effect on the life-cycle cost, and although the objective is to compare alternatives – the same rate being used for each – an error in the rate might affect their ranking in some circumstances. The rate to use will in fact vary depending upon the building owner. He either has to borrow money to finance his building or sacrifice the alternative use of his own money. The applicable discount rate should therefore be that rate of interest which he could obtain by investing his funds in an investment of equivalent risk. It will not necessarily be the interest rate for the mortgage on the building, and in the case of a developer putting up an income-producing building it could be the internal rate of return he expects on his development. In other cases it will have to be found in discussions with the building owner and may depend on the average internal rate of return of his business, his cost of borrowing and the anticipated rate of inflation. Or it may be a weighted average of the financial sources of his business, both debt and equity. Even the government, while it raises money principally by taxation, has to borrow and has a substantial national debt to be repaid, and any expenditure on building is not available for other purposes, so an appropriate discount rate can be found for government projects as well as for private projects.

The effect of the discount rate on costs can be seen in Tables 10 and 11 which show the present cost of annual expenditures of a thousand dollars and the present cost of future periodic expenditures of a thousand dollars respectively, for various interest rates over various periods of time. These tables show that an increase in the discount rate makes it less worthwhile to increase capital costs in order to save future annual or periodic costs. For example, if the life of a building is fifty years, with a discount rate of five per cent it is worth spending up to $18 256 in capital cost in order to save a thousand dollars in annual costs, while at twelve per cent it is only worth spending up to $8 304. The same effect would be indicated in an annual equivalent life-cycle cost study which would show a higher annual equivalent cost at a higher discount rate than at a lower rate. This means that the lower the discount rate, the more worthwhile it is to increase capital costs in order to save future costs. There is a tendency for high rates to encourage lower capital costs, and hence lower standards of construction because the resulting future operating and maintenance costs are heavily discounted, while low rates tend to encourage higher capital costs and better standards of construction.

Inflation

There are two types of inflation which might be considered for inclusion in a life-cycle cost analysis: General inflation in the economy as a whole, and escalation in construction costs. There is some disagreement about whether general inflation should be included in life-cycle costs, although most authorities agree that it should not be included. All life-cycle cost studies have to be based on assumptions about the future and it has been argued that as with a financial feasibility study, including general inflation as an additional assumption only increases the possibility of error. Also, interest rates tend to rise during periods of inflation so the two are likely to cancel each other out. Besides this, general inflation will affect all costs equally, and since life-cycle costing is an exercise in comparative costs, not in absolute costs, its inclusion should not affect the ranking of the alternatives.

Table 10 – Present cost of annual expenditures of $1,000

	5%	8%	10%	12%
10 years	7 722	6 710	6 145	5 650
20 years	12 462	9 818	8 514	7 469
30 years	15 372	11 258	9 427	8 055
40 years	17 159	11 925	9 779	8 244
50 years	18 256	12 233	9 915	8 304
60 years	18 929	12 377	9 967	8 324
70 years	19 343	12 443	9 987	8 330
80 years	19 596	12 474	9 995	8 332
90 years	19 752	12 488	9 998	8 333
100 years	19 848	12 494	9 999	8 333

Table 11 – Present cost of future periodic expenditures of $1,000

	5%	8%	10%	12%
10 years	613.91	463.19	385.54	321.97
20 years	376.89	214.55	148.64	103.67
30 years	231.38	99.38	57.31	33.38
40 years	142.05	46.03	22.09	10.75
50 years	87.20	21.32	8.52	3.46
60 years	53.54	9.88	3.28	1.11
70 years	32.87	4.57	1.27	0.36
80 years	20.18	2.12	0.49	0.12
90 years	12.39	0.98	0.19	0.04
100 years	7.60	0.45	0.07	0.01

If general inflation is considered to be a necessary ingredient in a life-cycle cost study it can be included by modifying the discount rate using the following formula:

$$MR = \left(\frac{1 + i}{1 + e} - 1\right) \times 100$$

where MR = the modified discount rate
and e = the rate of inflation.

This formula reduces the discount rate and a close approximation can be found merely by deducting the anticipated inflation rate from the discount rate to give the modified rate, which is usually close enough in practice. Thus, if a discount rate of ten per cent is to be used and inflation is expected to be four per cent per year, the modified discount rate will be 5.77% if calculated by the formula, and six per cent by simple deduction. This applies only to the costs which are included in a life-cycle cost study. If a positive cash flow such as income or salvage value has to be included the following formula must be used for modifying the discount rate of the positive cash flows only:

$$MR = [(1 + i)(1 + e) - 1] \times 100$$

This formula increases the discount rate and can also be approximated, this time by adding the anticipated inflation rate to the discount rate to give the modified rate. The reason why different formulae are needed for expenses and income is that inflation will cause future expenses to increase and future income to decrease in real terms. And, as can be seen from Tables 10 and 11, a reduction in the discount rate has the effect of increasing the present value of future costs while an increase in the discount rate reduces the present value of future income.

Although general inflation is not usually included in life-cycle cost studies, escalation in construction costs or in maintenance or operating costs frequently should be included, despite the difficulty of making these predictions. If, for example, two heating systems are being compared and there are indications that the fuel for one is likely to increase in cost at a higher rate than for the other, there is a strong case to be presented for making allowance for this price differential in the life-cycle costs. Similarly, when two materials with different lives are being compared and replacement costs will be involved, price escalation should be included, not only because of any difference in the rate of escalation but also because escalation will affect the comparative costs of replacement due to the difference in timing.

Example 9: What will be the life-cycle costs of the two roofs in Example 1 if the rate of escalation in construction costs is expected to be five per cent per annum?

Answer: Escalation can be allowed for in three ways: by using the formula given earlier to modify the discount rate, by deducting the escalation rate from the discount rate, or by allowing for escalation first and then discounting the result as follows:

	Roof A	Roof B
Initial cost	$2 000	$3 500

Repair at the end of 10 years:

Escalation: $200 × A
         = $200 × 1.62889 = $326
Discounted: $326 × V

CONSTRUCTION ESTIMATING AND COSTING

= $326 × 0.38554		126	—

Replacement at the end of 20 years
Escalation: = $2 000 × A
= $2 000 × 2.65330
= $5 307
Discounted: $5 307 × V
= $5 307 × 0.14864   789   —

Repair at the end of 20 years:
Escalation: = $200 × A
= $200 × 2.65330 = $531
Discounted: $531 × V
= $531 × 0.14864 =   —   79

Repair at the end of 30 years:
Escalation: = $200 × A
= $200 × 4.32194 = $864
Discounted: $864 × V
= $864 × 0.05731 =   50   —

Total Present Cost $2 965 $3 579

Including escalation has made no difference to the ranking of the alternatives and Roof A is still the more economical.

This example shows how allowing for escalation can increase the number of calculations if only discount tables are available and complete accuracy is required. It is bad enough when only periodic costs are involved, as in the example, but when annual expenses have to be considered a cash flow schedule has to be set up. The schedule needs to show the escalated annual cost each year, and then each must be discounted back to present values in a manner similar to that described in the last chapter when taxes have to be included in the calculation for the internal rate of return. Using the formula or simply deducting the escalation rate from the discount rate greatly simplifies the calculations. The life-cycle costs in this example would then be calculated in the same manner as in Example 1 except that the modified rate would be substituted for the actual discount rate. The use of the formula is considerably easier if a calculator with financial functions or a computer is available since the modified rate in Example 9 is actually 4.76% and the present value calculations would be difficult to make without this assistance.

Time

Time is another factor which has to be considered when studying life-cycle costs. It too can be considered under two headings: the life of the building itself, and the lives of the materials and components which go into it.

The life of a building very often depends not so much on its physical life as on its economic or functional life. Many buildings are constructed which could be expected to last many hundreds of years given the quality of the materials of which they are made, yet because of economic factors or changes in technology they become obsolete and have to be demolished within fifty years or less. How soon they become obsolete will depend very much on the type of building and its location, and to some extent on its architectural merit. Obsolescence may be due to a financial factor, as for example where it is found to be more profitable to demolish the building because the owner has found a better use for the site, or to functional obsolescence where the owner has found it to be cheaper to demolish and rebuild than to try to adapt the building to meet his changing requirements. This makes it difficult to forecast the probable life of a building although a realistic estimate is needed since it can affect the outcome of the life-cycle costs.

This is shown by reference again to Tables 10 and 11. If a building can be expected to have a long life it is more worthwhile increasing the capital cost in order to reduce future costs. At a five per cent discount rate it is worth spending up to $18 256 in capital cost in order to save future annual costs of a thousand dollars if the life of the building can be expected to be fifty years. On the other hand, if the building can only be expected to last for twenty years it is only worth spending up to $12 462 to make the same saving. This would also be reflected in the annual equivalent costs which would be higher for a short-life building than for a long-life building. Essentially it means that a building with a short life should have less spent on it in capital cost than a building with a longer life even though this may mean spending more on operating and maintenance costs. However, manufacturers of building materials and components do not always differentiate between those to be used in a short-life building and those to be used in a long-life building. Thus a building intended to have only a short life is not proportionately less expensive in capital cost than one intended to have a longer life; a quick turnover in buildings is therefore generally uneconomical.

Another conclusion which can be drawn from Tables 10 and 11 is that after a certain period of time the life of the building has only a small effect on the present value of future expenditures. At a five per cent discount rate this occurs after about sixty years, beyond which time an increase in the assumed life of the building will have only a minimal effect. At eight per cent it occurs after about fifty years, at ten per cent after forty years, and at twelve per cent after thirty years. One result of this is that any error in predicting future costs will cause a greater error in the life-cycle cost if the life of the building is underestimated than if it is overestimated, and underestimating the life of the building increases the probability of reaching a wrong conclusion. However, the shorter the assumed life of the building the less effect any error in the discount rate will have on the present value. With a life of ten years, the saving of an annual expenditure of a thousand dollars can be accomplished with a capital cost of $7 722 if the discount rate is taken to be five per cent, and $5 650 if the rate is taken to be twelve per cent, a difference of $2 072. With a life of fifty years the capital costs need to be $18 256 and $8 304 respectively, a difference of $9 951. This means that the accuracy of the discount rate is more important for a long-life building than it is for a short-life building.

The other aspect of time, that of the life of the materials or components which are incorporated in the building, can be equally difficult to predict. The lives of many materials and components are less than the life of the building, either because it is cheaper to replace than to repair them, or because they are not capable of being repaired, and they may be replaced several times during the building's lifetime. Little accurate information is available on the expected lives of materials and components, partly because their lives are dependent to some extent on the maintenance policy of the building owner, and partly because the amount of use they receive in the building will affect their longevity. A vinyl-asbestos tile floor, for example, will last longer in a building which is well maintained, particularly if it is located in an area which does not receive much wear, than it will in a building which is poorly maintained and subject to heavy traffic. Enquiries directed to the manufacturers of materials and components are likely to elicit optimistic responses which would lead to unrealistic estimates of a product's durability. A more fruitful source could be property managers and building owners who have had long experience in building maintenance. The problem of predicting the life of materials can be compounded when the life-cycle cost of a new material is required. In this case, since no data on past performance will be available, the life of the material will have to be assumed, or alternatively a calculation must be made of the life necessary for the life-cycle cost of the new material to equal that of the material with which it is being compared, and a decision made on the probability of this life expectancy being achieved.

Costs

Because life-cycle costing can be used to help members of the design team and the building owner to make rational decisions over a wide range of variables, from a choice between two or more building materials to a choice between two or more overall building designs, the amount of available cost information must be equally wide. Although the cost information usually has to be based on historical data when it is used in a life-cycle cost study it needs to relate to present and future costs, because decisions can only be made about present and future expenditures and the values they can create. Once money has been spent and the building has been constructed it means that all the decisions have been made and they cannot be recalled.

LIFE-CYCLE COSTING

The costs required for life-cycle costing can be broken down into a number of categories which, because of their application, come under three major headings: capital costs, annual costs, and periodic or cyclical costs.

Capital costs are the easiest to estimate. The information is usually readily available, and comparisons of materials, systems or buildings on the basis of capital costs are frequently made and are well understood. However, costs not directly related to the cost of labour and materials being used in the building may sometimes be required, such as the cost of land when a comparison is being made between the cost of a high-rise structure and the cost of a low-rise structure which would require additional land. If construction management is under consideration the effect of the reduction in design and construction time and any resulting difference in construction cost and design fees would be assessed and then compared with the value to the building owner of obtaining earlier occupancy.

In a financial feasibility study the annual costs are usually considered together under a heading of "Operating Costs" as was done in Chapter 7, but for the purpose of life-cycle costing it is often useful to separate them as follows:

1. MAINTENANCE COSTS

These are usually defined as those costs which are required in order to maintain the building as far as possible in its original condition. They would include the cost of labour and supplies for daily cleaning or the cost of a maintenance contract if it is to be done by an outside organization, a service contract for the servicing of elevators or other equipment, plus minor repairs and upkeep to the building fabric such as the replacement of electric light bulbs or tubes. Maintenance costs will depend to a large extent on the type of building, public buildings being likely to have higher maintenance costs than buildings in the private sector. They will also depend on the maintenance policy of the building owner. Some owners have a policy of planned maintenance in which maintenance tasks are carried out according to a set schedule. This can result in a better use of maintenance staff, less disturbance to the building occupants as maintenance operations can be more conveniently scheduled, a reduction in the need for stand-by equipment because there is less chance of failure in the main equipment and a more reliable service from the building. Planned maintenance, on the other hand, may also be comparatively more costly since there will be a greater amount of routine servicing and probably more frequent replacement of minor components, often before they have reached the end of their lives. However, planned maintenance can also mean that less expensive equipment might be considered for the building because it will be well maintained.

Estimating the cost of maintenance should be easier when the building owner adopts a policy of planned maintenance than it would be if he relies on remedial maintenance which, apart from routine cleaning, requires that maintenance work only be done when a failure indicates that it is required. In this case the selection of more expensive equipment which is less subject to breakdown should certainly be considered. It might also be noted that even a building owner with a policy of planned maintenance will tend to spend less on maintenance when there is a downturn in the economy.

Most owners of existing buildings have records of their maintenance costs, although these are compiled for accounting purposes and may not be in a form which can provide immediate data for a life-cycle cost study. Even so, the costs which are available can give some indication of the magnitude of maintenance costs for specific buildings, and further analysis can break the costs down into a more suitable form, particularly if they can be discussed with the owner or his building manager. If possible the analysis should go further than revealing the maintenance costs of individual materials and equipment. Ideally they should show the man-hours required for various maintenance operations together with the supplies and equipment required, much as is done when recording data for construction estimating, thereby allowing costs to be readily updated when necessary. If a building manager has already been appointed when the life-cycle cost studies are being prepared he can provide invaluable information on the probable maintenance and other anticipated costs and also offer useful suggestions on how these costs can be reduced.

2. OPERATING COSTS

These are costs which are directly related to the operation of the building such as the cost of fuel, electric power, water, security personnel and the like. They are usually easier to estimate than maintenance costs although some uncertainty can exist in predicting future costs, as was amply demonstrated when the cost of fuels for heating systems rose far higher than anyone could have predicted. As has already been mentioned, if there is a probability that there will be a variation in the difference between operating costs, such as the cost of one fuel increasing at a higher rate than another, this should be allowed for in a life-cycle cost study. However, allowing for escalation in operating costs is only likely to change the ranking between alternatives if the operating costs are large in relation to the capital costs, if the difference between the operating costs is likely to increase substantially, and if the difference is likely to take place in the near future.

3. USER COSTS

These are costs relating to the users of a building such as the cost involved in production in a manufacturing plant, or the cost of medical staff, equipment and supplies in a hospital. Life-cycle cost studies which include user costs

LIFE-CYCLE COSTING

are usually concerned with alternative planning solutions to see which provides the most efficient working environment within the building, or with comparisons between the cost of purchasing equipment and the saving in staff which might result from its use. User costs are therefore customarily only required for buildings which have a specific function and the costs are not usually difficult to obtain.

4. OTHER ANNUAL COSTS

These might include the cost of financing when a decision needs to be made about whether to delay construction until an anticipated reduction in mortgage interest rates takes place. Another example might be the cost of insurance when a decision is needed on the comparative costs of two types of construction, one of which will provide a better fire insurance rating. Although income is not usually thought of as an expense, and is not normally included in life-cycle costs, there may be occasions when the choice between two materials or systems will have an effect on the income potential of a building, in which case the difference in income would have to be allowed for.

Annual costs can represent a major proportion of the total costs of a building over its life, very often equalling if not exceeding the capital cost. The actual ratio between capital and annual costs will vary between building types, and within building types it will vary from building to building depending upon how well they have been designed and built, and how well they are managed.

Periodic costs would include the following:

1. REPLACEMENT COSTS

These are the costs of replacing those parts of the building such as finishes and equipment whose lives are shorter than that of the building. Redecorating can be considered as a replacement cost since it is normally a periodic cost and, in effect, it replaces the original decorations. Replacement costs of equipment will apply both to the equipment itself and to its component parts. When making a decision in the design stage between one piece of equipment which has a high capital cost and a long life and another which has a lower capital cost but a shorter anticipated life, the decision can depend on more than just the estimated cost of replacement. It may also depend on the anticipated life of the building and whether, in fact, the equipment with the lower capital cost may last as long as the building. The decision may also be influenced by changes which are likely to take place in the function of the building that could make it necessary to replace the equipment in a comparatively short time anyway, and whether there is a possibility that technological improvements will make the equipment obsolete before it reaches the end of its physical life. These factors may make the choice of the less expensive equipment the better decision.

A decision on the cost of replacing a defective component part is more likely to be needed during the life of a building than it is in the design stage of a new building. The available choices then might be to replace the component with an identical part; to replace it with an improved model which may be more expensive in capital cost but which is either more efficient, has a greater life expectancy, has lower running costs, or is a combination of all three; or just to repair the defective part. A decision on which course is best can be made if life-cycle costs are calculated for each.

The most common materials which have to be replaced during a building's life are roof finishes and interior finishes, particularly floor finishes, although there can be many others including windows, doors, parapet walls, railings and built-in fittings and fixtures. When considering the replacement cost of materials such as floor finishes, the cost of disturbance, if any, to the occupants of the building while the replacement is being made should also be included. As with components, a decision may be needed whether to replace or to repair.

2. REPAIRS

These are the costs of repairing those parts of the building which become defective. Repointing brick walls or recaulking windows and doors are periodic expenditures which are fairly predictable and which can be considered as repairs. The type and frequency of other repairs, however, can be almost impossible to forecast, although their cost may not necessarily be difficult to predict. As with most other expenses incurred during the life of a building, so much depends on the way the building is used and maintained, but in the case of repairs it can also depend on how the building is designed and constructed. Poor detailing, bad workmanship or faulty materials can result in repairs having to be made much sooner than was anticipated, and this is completely unpredictable. It is worth noting that repairs are required when there has been a failure, and in the past large capital sums have been spent in attempting to avoid failures which might never occur or which would have been less expensive to repair. A life-cycle cost analysis could have revealed this unnecessary capital expense.

3. ALTERATION COSTS

These are the costs of making alterations in the building after it has been occupied for some time. Alterations are not usually initiated unless there is an evident economic gain to be made, either in making the operations within the building more efficient or in reducing the overall running costs of the building. Sometimes, however, outside factors can influence the decision, as for example when a change takes place in government policy with respect to the operation of hospitals or changes in population trends affect the need for schools.

Alteration work is always more expensive than work done at the time the building is being constructed. It also has to be remembered that the cost of alterations should include the cost to the building's occupants of rescheduling their activities while the work is in progress. In some instances this can be more than the actual cost of construction, involving a loss of production or the need for overtime in a manufacturing plant; the loss of income in a rental building, together with the cost of providing temporary alternative accommodation elsewhere if necessary; and the loss of amenities in a hospital or school. Evaluating the loss of amenities may pose some problems, but nevertheless it should be costed and allowed for in a life-cycle cost study. In addition there will be the costs of moving and protecting furniture and fittings while the alteration work is being carried out.

Because of the cost of alteration work, and the difficulty of predicting when and to what extent it may be required, provision is often made when the building is being constructed to make it more amenable to future alterations. A simple example of this is the use of movable partitions in place of solid partitions. Another example is providing increased structural requirements to give more flexibility for future alterations. It is extremely expensive to make structural alterations in an existing building, and increased column spacing, greater floor-bearing capacity and increased storey heights can be incorporated at much less expense at the time of construction than if they have to be provided later. In recent years the inclusion of interstitial space in hospitals has also been advocated as a means of reducing future alteration costs. It is usually combined with long spans which give greater flexibility to the working floors as well as providing additional space separated from the working floors for material handling equipment and the electrical and mechanical services which can then be altered without disturbing the occupants of the building. Making these structural provisions for future alterations will undoubtedly add to the capital cost. The structural costs themselves will increase and so will the quantity and cost of exterior cladding, partitions and stairs since making these structural provisions usually entails increasing the storey heights. In addition, the operating costs of the building will be higher because there will be a larger building volume to be heated and cooled, and interstitial space will require lighting.

It is difficult to justify spending large sums in capital cost to save future costs which may never be incurred, at least not to the extent which warrants the increased capital expenditure. However, a life-cycle cost study will help to show whether or not it is warranted. In this case it might take the form of estimating the premium in capital costs for providing the additional requirements, and discounting this sum over the life of the building to give an annual equivalent. Any additional operating costs would then be added to show a total annual cost, and the decision would be based on whether the annual saving in alteration costs is likely to be more or less than this amount. For example, a premium of $2 000 000 in capital cost for providing interstitial space has an annual equivalent cost of $167 700 over a forty-year period at a discount rate of eight per cent. If the expected premium in operating costs is $8 000 a year the total annual cost will be $175 700. If the extra cost of making alterations each year in a scheme without interstitial space can reasonably be expected to exceed this amount, then the interstitial space is justified. Alternatively the study might take the form of finding the period of time within which the capital cost of making provision for future requirements equals the cost of making alterations in the future. The further in the future that alterations are likely to be needed, the less advantage there is in making provision for them at the time of construction.

Example 10: The owner of a factory which is currently being designed believes that changes in production techniques may require certain alterations to the building at some time in the future. Making provision for these future requirements now will cost an additional $75 000, while making the alterations in the future is estimated to cost $130 000 at current prices. When will the break-even point be, assuming a discount rate of 9% and a construction escalation rate of 5%?

Answer: The modified discount rate incorporating escalation will be 9% less 5% which equals 4%. The costs will break even when the capital cost and the present value of the future cost equal each other, that is when:

$$\$130\ 000 \times V = \$75\ 000$$

$$\text{or } V = \frac{75\ 000}{130\ 000}$$

$$= 0.57692$$

Reference to PRESENT WORTH OF ONE DOLLAR tables shows that at a discount rate of 4% this occurs in about fourteen years. If the owner thinks the alterations will be required within the next fourteen years the increased capital expenditure is justified.

This calculation becomes a little more complicated if more than a single future expenditure is involved.

Example 11: What will be the break-even point in Example 10 if, in making provision for future requirements the operating costs increase by $1 000 a year?

Answer: The break-even point will be when:

$$\$130\ 000 \times V = \$75\ 000 + \$1\ 000 \times YP$$

$$\text{or } \frac{\$130\ 000 \times V}{\$75\ 000 + \$1\ 000 \times YP} = 1$$

LIFE-CYCLE COSTING

As with the calculation of the internal rate of return in Chapter 10 this can be solved by interpolation using two target break-even points, in this case 10 and 15 years.

At 10 years:
$$\frac{\$130\,000 \times 0.67556}{\$75\,000 + \$1\,000 \times 8.11090}$$

$$= \frac{87\,823}{83\,111}$$

$$= 1.05670$$

At 15 years:
$$\frac{\$130\,000 \times 0.55526}{\$75\,000 + \$1\,000 \times 11.11839}$$

$$= \frac{72\,184}{86\,118}$$

$$= 0.83820$$

Break-even point = Lower life + $\frac{HR - 1.0}{HR - LR}$ (Higher life – Lower life)

$$= 10 + \frac{1.0567 - 1.0}{1.0567 - 0.8382} \times 5$$

$$= 10 + \frac{0.0567}{0.2185} \times 5$$

$$= 10 + 1.29748$$

$$= 11.29748 \text{ years.}$$

With the additional operating costs the increased capital expenditure is now justified only if the owner thinks the alterations will be required within a little over eleven years.

In practice it is hardly necessary to go into the fine detail shown in the last example. A glance at the ratio produced for ten years shows that the break-even point won't be much beyond ten years and this is probably close enough.

Another aspect of alteration costs is that of having to decide whether to renovate an existing building or whether to demolish it and rebuild. Money spent on renovating existing buildings — which may only have a limited life anyway — is not then available for constructing new buildings, and despite the renovations an existing building may lack the amenities of a new one and have higher maintenance and operating costs. A life-cycle cost study in this instance could take the form of a feasibility study showing the present value of the renovated building for comparison with the present value of a new building. Although this will help make an economic decision about whether or not to renovate, the decision sometimes is not entirely an economic one, but is influenced by community interests and pressures, architectural merit and other intangible factors.

4. SALVAGE

This, like income, can be considered as a negative cost. Some materials and equipment may have a salvage value at the end of their life and this should be included in a life-cycle cost study when appropriate. Salvage will also help offset the cost of demolition. Although demolition is generally thought of as being the cost of demolishing an existing building prior to the erection of a new one and would therefore be a capital cost rather than a periodic cost, there will be occasions when demolition and salvage have to be considered as periodic costs occurring once at the end of the building's life.

The capital, annual and periodic costs just described cover most of the tangible costs which might be needed for life-cycle costing. There are, however, other costs which may be needed and which are much harder to estimate. These might relate to the appearance of the building, to the standards of comfort and safety of those working in it, and to the effect of the building on the surrounding community. Appearance has already been mentioned as a factor which will depend upon the building owner's objectives, and assessing any difference in potential income because of the building's appearance should not be difficult. However assessing the ease of raising capital or the advertising value which will accrue to a building of enhanced prestige is more problematical. Improving the standards of comfort and safety for the building's occupants might be reflected in improved staff morale and could be related to a possible reduction in the cost of salaries and a lower staff turnover. The effect of the building on the community is rarely considered by most building owners unless it reflects directly on the profitability of the building or the business which is being carried on within it. Such matters as improved access and parking, superior landscaping, and noise abatement and anti-pollution devices beyond those required by the law might all be considered as reflecting a building owner's regard for the community, but placing a value on them is extremely difficult.

Taxation

Taxation is not a factor which needs to be considered in life-cycle cost studies for public buildings such as government offices, hospitals and schools. Whether it need be considered in a study for a private owner is debatable. From the last chapter, which described how taxation can be incorporated into the financial feasibility study of a development project, it can be seen that the amount of tax a building owner pays annually is largely a matter of his fiscal policy, combined with his method of financing and the income potential of his building, and subject therefore to a certain amount of manipulation. For this reason taxes are liable to fluctuate widely and will have little direct bearing on the type or cost of individual materials, components or systems. Unless there are obvious tax advantages in the choice of a material, component or system — which is highly unlikely — or the life-cycle cost studies are to be performed on total building designs, taxation rarely needs to be considered in life-cycle cost studies.

Error Analysis

Life-cycle costing involves attempting to forecast the future, usually well into the future, and it must be based on numerous assumptions. Assumptions can sometimes prove to be wrong so it is useful to be able to test them to see whether any variations will affect the validity of the results.

Assumptions have to be made about costs, both capital and future costs. It is usually considered that any error in costs will be self-cancelling because life-cycle costing is concerned with comparative costs, not with absolute costs, but while this is true in many circumstances, and complete accuracy in the cost is not usually essential, there can be occasions when an error in the costs can affect the ranking. If the costs being compared both err in the same direction, that is they have both been overestimated or underestimated; or if the higher cost has been underestimated; or the lower cost has been overestimated, there will be no change in the ranking. However, if the costs being compared both err in opposite directions; or the higher cost has been overestimated; or the lower cost has been underestimated, the ranking can be changed. If, for example, two walls have capital costs of $100 000 and $110 000 the change in ranking can take place in the following circumstances:

(a) The higher cost has been overestimated and the lower cost has been underestimated by about five per cent in both cases:
$110\ 000 - 5\% = \$104\ 500$
$100\ 000 + 5\% = \$105\ 000$

The actual percentage when both costs will break even can be found as follows:

$$100(1+x) = 110(1-x)$$
$$100 + 100x = 110 - 110x$$
$$100x + 110x = 110 - 100$$
$$210x = 10$$
$$x = \frac{10}{210}$$

$$= 0.4762 \quad \text{or say } 4.76\%$$

(b) The higher cost has been overestimated by a little over nine per cent:
$$\$110\,000 - 9.09\% = \$100\,001$$
(c) The lower cost has been underestimated by ten per cent:
$$\$100\,000 + 10\% = \$110\,000.$$

If these percentage errors are considered to be likely, the capital costs of the two walls can be considered to have no significant effect on their life-cycle costs. The possibility of changes in future costs can be assessed in a similar manner. If two heating systems have operating costs of $12 500 and $9 500 per year, the more expensive system would have to reduce its operating costs by twenty-four per cent to make it comparable with the less expensive system, and the probability of this happening would have to be judged in making a decision.

The effects of changes in the discount rate and in the assumed life of the building have already been mentioned. The extent to which these will affect the ranking in life-cycle cost studies will depend to a large degree on the ratio between the capital costs and the future costs, particularly annual costs, and special care is needed when one design has a low capital cost and high annual costs while the alternative has the reverse. It is often worthwhile carrying out sensitivity analyses to see what effect changes in the costs, the discount rate and the assumed lives of the building or the materials or components will have on the results.

Example 11: What will be the effect on the comparison made in Example 3 if Floor A needs replacing only every fifteen years and the discount rate is increased to ten per cent?

Answer:

	Floor A	Floor B
Initial cost	$10 000	$15 000
Replacement costs:		
V @ 15 years = 0.23939		
30 years = 0.05731		
0.29670		
Floor A: $10 000 × 0.29670 =	2 967	—
Floor B: $15 000 × 0.29670 =	—	4 451
Maintenance:		
Floor A: $1 200 × YP, 40 years @ 10%		
= $1 200 × 9.77905 =	11 735	—
Floor B: $800 × YP, 40 years @ 10%		
= $800 × 9.77905	—	7 823
Total present cost	$24 702	$27 274

These changes to the expected life of Floor A and the discount rate have reversed the order found in Example 3, and Floor A is now the less expensive.

This example shows how changes in two of the variables can affect the ranking of the alternatives. The number of changes which could be tested is limitless so it is advisable to restrict them only to those assumptions which are felt to be subject to variation. Generally speaking, the testing should be restricted to the outer limits of the possible variations, unless a computer is available, in which case a wide range of variations in discount rates, lives, costs and escalation rates can be tested and, if necessary, plotted on graphs to visually demonstrate how they affect the ranking of the alternatives.

Conclusion

It has been said that life-cycle costing is not appropriate if the building owner expects to sell the building after a few years because he should then only be concerned with capital costs since the new owner will be responsible for the maintenance and operating costs. A similar argument is made for publicly owned buildings where one government department is responsible for the capital costs and another for the maintenance and operating costs. However, the objective of life-cycle costing is to help give the building owner the best value for his money, and this means that the value of the building is likely to be enhanced as a result. In the case of the private owner this will be reflected in the price the new owner is prepared to pay for the building, so both the original and the new owner will reap the benefits of the life-cycle cost studies. In the case of the public owner no distinction should be made between which government department is spending the money; it is still public money and should be spent wisely, and life-cycle cost studies will help ensure that this is done.

Judgement is needed when making decisions based on life-cycle cost studies. Overexpenditures on capital costs in order to save future maintenance and operating costs can result in increased financing costs which can sometimes inhibit a building owner's future actions to nearly the same degree as spending too little on capital costs and as a result burdening him with high maintenance and operating costs.

All the previous examples in this chapter have given comparatively simple illustrations of the way in which life-cycle cost studies might be performed. The next, and final, example gives a more complex illustration incorporating many of the factors which have been discussed in this chapter.

Example 12: Which of the following two exteriors walls is the more economical given the following data all of which relates directly to the exterior walls:

	Wall Type A	Wall Type B
Initial cost of the wall, per m^2	$ 56.00	$ 65.00
Total area of the wall, m^2	12 300	12 300

Salvage value, per m² of wall	$ 4.00	$ 8.50
Cleaning exterior, per m² of wall:		
Every 25 years	$ 1.20	—
Every 6 months	—	$ 0.50
Painting exterior, per m² of wall:		
Every 25 years	$ 12.00	—
Caulking exterior, per m² of wall:		
Every 10 years	—	$ 1.40
Painting interior, per m² of wall:		
Every 4 years	$ 2.10	—
Fire insurance, per $100 building cost:		
Building, per annum	$ 0.08	$ 0.10
Contents, per annum	$ 0.20	$ 0.25
Total gross floor area of the building, m²	25 600	25 600
Cost per m² gross for the building	$ 325.00	$ 332.50
Total capital cost of the foundations	$ 763 000	$ 748 000
Total capital cost of electrical	$1 275 000	$1 150 000
Hydro costs, per annum	$ 90 200	$ 85 400
Total capital cost of heating and air conditioning	$1 950 000	$2 150 000
Fuel and maintenance costs per annum	$ 40 200	$ 46 800
Total net floor area of the building, m²	$ 19 340	$ 19 500
Rental rate, per m²	$ 105.00	$ 105.00

Assume the life of the building to be 40 years, and the discount rate to be 11%.

Answer:

Wall Type A:

Initial cost of the wall: 12 300 m² @ $56.00 =	$ 688 800
Premium for foundations: $763 000 − $748 000 =	15 000
Premium for electrical: $1 275 000 − $1 150 000 =	125 000
Premium for hydro costs: ($90 200 − $85 400) × YP = $4 800 × 8.95105 =	42 965
Cleaning exterior: 12 300 m² @ $1.20 = 14 760	
Painting exterior: 12 300 m² @ $12.00 = 147 600	
$162 360	

$162 360 × V (25 years) = $162 360 × 0.07361 =   11 951

Painting interior: 12 300 m² × $2.10 = $25 830

V for 4 years = 0.65873
       8 years = 0.43393
      12 years = 0.28584
      16 years = 0.18829

LIFE-CYCLE COSTING

$$
\begin{aligned}
20 \text{ years} &= 0.12403 \\
24 \text{ years} &= 0.08170 \\
28 \text{ years} &= 0.05382 \\
32 \text{ years} &= 0.03545 \\
36 \text{ years} &= \underline{0.02335} \\
& \underline{1.88514}
\end{aligned}
$$

$25 830 × V = $25 830 × 1.88514 =	48 693
	$ 932 409
Less Salvage value: 12 300 m² @ $4.00 = $ 49 200	
$49 200 × V = $49 200 × 0.01538	757 Cr
Total present cost of Wall Type A	$ 931 652

Wall Type B

Initial cost of the wall: 12 300 m² @ $65.00	$799 500
Premium for heating and air conditioning:	
$2 150 000 − $1 950 000 =	200 000
Premium for design fees:	
8% of $192 000	15 360

Premium for fire insurance:
 Type B − Building cost: 25 600 m² @ $332.50
  = $8 512 000
 Annual premium = 85 120 @ ($0.10 + $0.25) = $29 792
 Type A − Building cost: 25 600 m² @ $325.00
  = $8 320 000

Annual premium = 83 200 @ ($0.08 + $0.20) =	23 296	
Extra	$ 6 496	
$6 496 × YP = $6 496 × 8.95105 =		58 146
Cleaning exterior: $0.50 every 6 months = $1.00 per year		
12 300 m²@ $1.00 = $12 300		
$12 300 × YP = $12 300 × 8.95105 =		110 098
Premium for fuel and maintenance costs:		
($46 800 − $40 200) × YP		
= $6 600 × 8.95105 =		59 077

Caulking exterior: 12 300 m² @ $1.40 = $17 220

$$
\begin{aligned}
V \text{ for 10 years} &= 0.35218 \\
V \text{ for 20 years} &= 0.12403 \\
V \text{ for 30 years} &= \underline{0.04368} \\
& \underline{0.51989}
\end{aligned}
$$

$17 220 × V = $17 220 × 0.51989 =	8 953

Less increased income: (19 500 m² − 19 340 m²)
@ $105.00 = $16 800
$16 800 × YP = $16 800 × 8.95105 =   150 378 Cr
Less Salvage value: 12 300 m² @ $8.50 = $104 550
$104 550 × V = $104 550 × .01538 =   1 608 Cr

Total present cost of Wall Type B $ 1 099 148

WALL A therefore appears to be the more economical, both in capital and in life-cycle costs.

The credit for increased income shown for WALL B is due to the difference in thickness of the two walls, giving the building a greater net floor area with WALL B than with WALL A.

Life-cycle costing is appropriate in a wide variety of situations when a building is being designed or renovated, or when replacements or repairs are being considered, and is applicable whenever the capital costs, running costs and lives of alternatives differ.

APPENDIX A

Method of Measuring the Gross Floor Areas of Buildings

Adapted from Measurement of Buildings by Area and Volume, *published by the Canadian Institute of Quantity Surveyors, Suite 704A, 43 Eglinton Avenue East, Toronto, Ontario M4P 1A2.*

1. GENERAL REMARKS

The measurement of the gross floor area has been used by the construction industries of a number of countries throughout the world as a means of comparing buildings of similar type and construction.

One of the principal reasons for this comparison is to relate the costs of buildings of a similar type in terms of both location and time.

The intent of the procedures listed hereafter is primarily for the purpose of standardizing a method of measuring gross floor areas for both new and existing buildings.

It should be emphasized that the use of an area cost applied to the gross floor area of a building is, at best, only a means of checking an estimated cost found by more accurate methods. Particular attention should be given to those items listed under INCLUSIONS and EXCLUSIONS, together with any other necessary allowances and adjustments for non-standard units or conditions. Care should also be taken to ensure that previous unit costs have been obtained using measurements based on the following procedures.

GENERAL RULES

(a) Measure all usable portions of the building within the outside face of the exterior walls together with any additional items listed hereafter.
(b) The outside face of the exterior wall is defined as the outside face at each floor level, but excludes such horizontal features as projecting cornices, stone bands, etc.

(c) Exterior walls are defined as walls made of any permanent material, applied vertically or nearly vertically for the purpose of weatherproofing the building to acceptable use and occupancy standards.
(d) Sloping and stepped floors shall be measured flat on plan.
(e) In all cases measurements shall be taken in metres to two decimal places, and the results expressed in square metres to the nearest square metre.
(f) The rules for rounding shall be as described in the *Canadian Metric Practice Guide* published by the Canadian Standards Association, as follows:
 (i) When the first digit discarded is less than five, the last digit retained should not be changed.
 Example: 3.143 rounded to two decimal places = 3.14
 (ii) When the first digit discarded is greater than five, or if it is five followed by at least one digit other than zero, the last figure retained should be increased by one unit.
 Examples: 3.137 rounded to two decimal places = 3.14
 3.135 01 rounded to two decimal places = 3.14
 (iii) When the first digit discarded is exactly five, followed only by zeros, the last digit retained should be increased by one if it is odd, but no change made if it is even.
 Examples: 3.135 000 rounded to two decimal places = 3.14
 3.145 000 rounded to two decimal places = 3.14
(g) Where a new building adjoins an existing building, take measurements for the new building only.
(h) Measure each building separately where more than one building is involved and classify under its occupational usage. In the case of adjoining buildings with party walls, take measurements for each to the centre line of the wall between buildings.
(i) The gross floor area is defined as the total area for each building measured in accordance with these rules.

3. MEASUREMENT OF AREA (SQUARE METRES)
A. Method

Measure from outside face to outside face of exterior walls for the area of each floor, making no deductions for openings which occur within the floor area, except as noted hereinafter. Where balconies and mezzanine floors occur within the exterior walls of the building the actual area of these shall be measured and included. When the exterior walls are broken up with a large number of small projections (e.g. projecting columns) the measurements shall be taken to the mean outside face of the exterior wall.

B. General Notes

1. Make no deductions to the area for:

APPENDIX A

   (a) walls, partitions, etc.;
   (b) openings in floors for stairwells, escalators, elevators, ducts and other similar facilities;
   (c) pits, trenches, depressions occurring in the lowest floor which are open or have removable covers;
   (d) columns, piers or pilasters;
   (e) any other features within the confines of the exterior walls.
2. Where auditoriums, swimming pools, gymnasiums, foyers and the like extend through two or more floors they shall be included for the largest area, at one level only.

C. Inclusions

Include the following items in computing the gross floor area:
(a) crawl spaces with concrete floors;
(b) basements and future basement areas where a concrete slab only is required for structural completion;
(c) floor areas which are structurally completed and where the finishing work will be executed at a later date;
(d) mezzanine floors;
(e) interstitial floors;
(f) tunnels, pits, trenches, etc., which have a covering roof slab or grating that is 2.00 m or more from the floor;
(g) rooms below grade or sidewalk (e.g. transformer rooms);
(h) dormers, bay windows and the like provided they extend vertically for the full floor height;
(i) penthouses;
(j) elevator machine floors within penthouses;
(k) connecting links or walkways provided they are enclosed;
(l) finished rooms in roofs and attics;
(m) attached or isolated garages above and/or below ground level;
(n) fully enclosed exterior staircases and fire escapes;
(o) fully enclosed porches.

D. Exclusions

Exclude the following items in computing the gross floor area. However if conditions warrant the inclusion of some of these items they shall be listed separately from the gross floor area:
(a) crawl spaces which do not have concrete floors;
(b) tunnels, pits, trenches, etc., with less than 2.00 m head room;
(c) exterior balconies;
(d) canopies;
(e) projections beyond the face of the exterior walls;

(f) doghouses on roofs;
(g) areaways;
(h) unenclosed connecting links;
(i) covered walkways with no side walls;
(j) unfinished roof and attic areas;
(k) carports;
(l) unenclosed exterior staircases and fire escapes;
(m) isolated chimneys and that portion of chimneys above the roof line;
(n) interior open court yards, light wells, atrium voids and the like;
(o) unenclosed porches;
(p) exterior steps and landings;
(q) exterior paving, patios and terraces;
(r) unenclosed areas which are roofed over;
(s) enclosed areas which are not roofed over;
(t) roof overhangs and cornices.

APPENDIX B

Definition of Elements

Adapted from a list of Elements prepared by the Canadian Institute of Quantity Surveyors, Suite 704A, 43 Eglinton Avenue East, Toronto, Ontario, M4P 1A2.

Preamble

(a) An Element is defined as a major component common to most buildings, usually fulfilling the same function irrespective of its design, specification or construction.

(b) Where mention is made of different types of construction and materials having to be kept separate within Elements this applies to the back-up material needed to prepare a cost analysis and to the documentation needed to prepare an estimate. The cost analysis itself normally need only show a total quantity (where appropriate), a ratio, a total cost, a unit rate and a cost per square metre of the gross floor area for each Element. The gross floor area is to be measured in accordance with the rules laid down in Appendix A.

(c) Because it is difficult to make hard and fast rules for defining exactly the extent of each Element which will apply to every building the definitions should be taken as definitions of principle. Where any departures seem necessary they should be clearly stated with the analysis.

(d) All Elements in a cost analysis should be shown in the sequence given here. If no cost is attributable to an Element, "NONE" should be written in the Quantity column and a dash put in the Amount column.

(e) Whenever possible, copies of plans and elevations (preferably dimensioned and reduced to letter size) should be attached to the cost analysis together with any suitable photographs and/or additional explanatory material.

CONSTRUCTION ESTIMATING AND COSTING

List of Elements

1. SUBSTRUCTURE
(a) Normal Foundations

Includes: Excavation, concrete or masonry for wall and column footings, column caps, grade beams, base plates, anchor bolts, foundation walls to the top of the lowest floor construction and weeping tile.
Excludes: Lowest floor construction, excavation for basements, basement walls and waterproofing, caissons and piles.
Measure the gross area in square metres to the outside face of perimeter walls or to the outer extent of foundations if there are column or similar foundations isolated beyond the perimeter walls. This area will usually be the gross area of the lowest floor and is sometimes known as the footprint of the building.

(b) Basement

Includes: The mass excavation and backfill required to construct a basement.
Excludes: Basement walls and waterproofing, and any work included in Element 1(a) Normal Foundations.
Measure the cube of the basement in cubic metres to the outside face of the perimeter walls and from the underside of the lowest floor construction to grade level.

(c) Special Foundations

Includes: Caissons, piling, extra cost of excavating in rock, special shoring, unusual dewatering and other special foundation conditions.
Measure rock excavation in cubic metres; shoring in square metres; and caissons and piling in metres keeping different types and sizes separate. Unusual dewatering is included as a lump sum allowance.

2. STRUCTURE
(a) Lowest Floor Construction

Includes: The floor slab, whether a slab on grade or a suspended slab over a crawl space (in the latter case it will include the columns and beams supporting the slab), fill under the slab, waterproofing and skim coat or vapour barrier to the slab, small sump pits, construction and expansion joints.
Excludes: Final finish to the slab and machinery or equipment bases.

APPENDIX B

Measure the gross area of the lowest floor slab in square metres to the outside face of the perimeter walls. (Note: This area is frequently the same as that for Element 1(a) Normal Foundations.)

(b) Upper Floor Construction

Includes: Columns, beams, slabs, floor joists, sub-floors and fireproofing for all suspended floors.
Excludes: Floor and ceiling finishes; and columns, beams, slabs, etc. for roof construction.
Measure the gross area of suspended floors in square metres to the outside face of the perimeter walls, keeping each type of construction separate. Make no deduction for openings. (Note: The gross area of Upper Floor Construction plus the gross area of Element 2(a) Lowest Floor Construction should equal the total gross floor area of the building.)

(c) Roof Construction

Includes: Columns, beams, slabs, rafters, purlins, trusses, roof boarding, etc. for roof construction.
Excludes: Roof finish, insulation, cant strips, flashings, roof lights, roof drains, eavestroughs and rainwater leaders.
Measure the gross area of the roof in square metres to the outside face of the perimeter walls, or if roof projects beyond the walls, to the farthest extent of the roof, keeping each type of construction separate. The area of sloping roofs are to be measured on the slope.

3. EXTERIOR CLADDING
(a) Roof Finish

Includes: Roof finish, insulation, cant strips, flashings, fascias, eaves soffits, barge boards, parapet walls above the roof level, copings and roof lights.
Measure the gross area of the roof finish as for Element 2(c) Roof Construction.

(b) Walls Below Ground Floor

Includes: Perimeter basement walls of reinforced concrete, brick, block, etc., including waterproofing, insulation and integral pilasters.
Excludes: All finishes (plaster, paint, panelling, etc.) to the interior face of the wall.

Measure the area of the walls in square metres from the top of the lowest floor construction to ground level. Keep different types of wall construction separate.

(c) Walls Above Ground Floor

Includes: Exterior walls of wood, stone, precast concrete, marble, face brick, curtain wall, etc.; brick, block, tile or concrete backup; insulation, vapour barriers, damp-proof courses, construction and expansion joints as well as applied finishes to the exterior face of the walls.

Excludes: All finishes (plaster, paint, panelling, etc.) to the interior face of the wall and parapet walls above the roof level.

Measure the gross area of the exterior walls in square metres along the general face of the walls from the top of the walls below ground floor, if any, or from ground level to the top of the roof finish if the roof is flat, or to the top of the wall if the roof is sloping. Ignore any extra perimeter caused by projecting isolated columns or recessed windows, etc., and deduct for windows and doors. Keep different types of wall construction separate.

(d) Windows

Includes: Steel, wood and aluminum sash; glazing; louvres; hardware; mullions; transoms; sills; stools; flyscreens; storm windows; lintels; precast concrete, stone, etc. surrounds; damp-proof courses; caulking; and all necessary painting or finishing.

Excludes: Curtain walls.

Measure the area of windows in square metres, keeping each type separate. (Note: Record for future reference the percentage of area of the windows to the gross area of Walls Above Ground Floor.)

(e) Exterior Doors and Screens

Includes: Wood, hollow metal, kalamein, steel, bronze and aluminum doors; roll-up or sliding shutters; store fronts; glazed entrance screens; revolving doors; decorative screens; metal and wood frames; subframes; sills; lintels; damp-proof courses; concrete, stone, etc. surrounds; caulking; electric or hand-operated opening devices; hardware; and all necessary painting and finishing.

Measure all glazed screens, store fronts and doors in square metres keeping each type separate.

APPENDIX B

(f) Balconies and Projections

Includes: Any item which, because of its existence, increases the area and cost of exterior cladding.
Examples are:
(1) Overhangs: If a floor projects beyond the floor below include the exposed soffit and insulation (but not the floor slab).
(2) Balconies: Include projecting balconies in their entirety. For recessed balconies include the cost of balcony railings, soffit finishes and insulation, but not the building wall, floor slabs or floor finishes. (Floor finishes are included in Element 3(a) Roof Finish)
(3) Canopies: Include projecting canopies attached to the building in their entirety.
(4) Sunshades: Include precast concrete, stone, wood, etc. sunshades which stand out beyond the face of the building.

General Comments on Exterior Cladding

(a) Record total cost of Elements 3(b) – (f) inclusive, and express as a price per square metre of the total area of vertical exterior cladding. This rate is very useful in comparing the costs of various sytems of enclosing space and can also be helpful in very early cost estimates when little or no information is available. Because of the wide range of materials available to enclose space this rate will vary considerably and must be used carefully in making early estimates. Record the ratio between the total area of vertical exterior cladding and total gross floor area as an indication of design efficiency.
(b) As it is sometimes difficult to distinguish between walls, doors, windows and screens, as in the case of curtain walls, certain Elements may have to be combined to avoid entirely arbitrary divisions.

4. INTERIOR PARTITIONS AND DOORS
(a) Permanent Partitions and Doors

Includes: Wood studs; metal studs; clay tile, concrete block, poured concrete or other solid walls or partitions; toilet partitions; and all interior doors located in permanent partitions, including hardware and all necessary painting or finishing of doors and frames.
Excludes: All finishes (plaster, gypsum board, paint, panelling, etc.) to walls or partitions.
Measure the gross area of all partitions in square metres and keep separate in different thicknesses and types of construction. Measure doors in square metres keeping each type separate.

(b) Movable Partitions and Doors

Includes: All movable, glazed or unglazed partitions, regardless of material, including doors; and folding or demountable partitions.
Measure movable partitions and doors in square metres keeping each type separate.

(c) Glazed Partitions and Doors

Includes: All permanent interior glazed partitions and doors (such as those forming vestibules at main entrances) glazed internal store fronts, and borrowed lights.
Measure glazed partitions and doors in square metres keeping each type separate.

General Comments on Interior Partitions and Doors

Record and express the total cost of Element 4 (a) - (c) inclusive as a price per square metre of all interior partitions and doors. This unit price is very useful for comparative purposes when considering the cost of this Element in various buildings. However, a wide range is possible because of the large number of materials available and the number of combinations of systems, and when using this unit rate to estimate projects in the very early stages great care must be taken to assess the many factors which are involved.

5. VERTICAL MOVEMENT
(a) Stairs

Includes: Treads, risers, stringers, landings, supporting framework, balustrades, handrails, soffits, steps, ladders, painting and other finishes, and floor framing around stair openings.
Excludes: Enclosing walls.
Measure the total length of risers in metres keeping each type of stair separate.

(b) Elevators and Escalators

Includes: Passenger and freight elevators, hoists, dumbwaiters and conveyors, including entrances, cabs and guide rails; escalator treads and balustrades, etc.; machinery and controls; and general contractor's work to provide bases, pits, fixings and floor framing around openings, etc.
Excludes: Hoistway enclosure walls and machine room power supply.

It is advisable to obtain quotations from manufacturers for this equipment and add general contractor's work required therefor. Record costs per elevator for different sizes, number of floors, speeds and degree of automatic control; and per escalator for complete installation, kept separate according to width, height, speed, etc., for calculating preliminary estimates for future projects. Because this is very specialized work this informaion should not be relied on too heavily except where very similar installations are being considered.

6. INTERIOR FINISHES
(a) Floor Finishes

Includes: All floor finishes including toppings, steel trowel or wood float finish together with bases, curbs, mat sinkages, frames and mats, construction joint cover plates, etc.
Measure floor finishes in square metres keeping each type separate.

(b) Ceiling Finishes

Includes: All ceiling finishes, cornices, bulkheads, etc.
Excludes: Special illuminated or heated ceilings.
Measure ceiling finishes and bulkheads in square metres keeping each type separate. Measure cornices in metres keeping each type separate. (Note: In the case of sloping, stepped or special ceilings measure the area on plan rather than the actual area for the quantity to be shown in a Cost Analysis.)

(c) Wall Finishes

Includes: All applied interior finishes to exterior walls and partitions, including grounds and strapping.
Excludes: Self-finished walls, the finish for which is included with the wall.
Measure wall finishes in square metres keeping each type separate.

General Comments on Interior Finishes:

Measuring interior finishes, particularly base and wall finishes, can be very time-consuming, and frequently their cost amounts to only a small percentage of the total construction cost. To save time the following is suggested as a quick method of measuring wall finishes for a preliminary estimate. Take once times the area of Elements 3(b) and 3(c) and twice times the area of Element 4(a). When added together this gives a gross area of interior wall surfaces, but since this usually exceeds the total area to be finished because of unfinished spaces and areas above ceilings, a percentage must be applied to reduce it to the total finished area. The percentage will vary with the type of building and each time a detailed estimate is made the percentage of finished wall surfaces to gross wall surfaces should be calculated for future use. As a general rule this usually works out to between 70 and 75 percent. As a further refinement to aid in calculating interior finishes when doing preliminary estimates, the percentage of each type of floor, ceiling and wall finish to the total finished area should be calculated and recorded when doing detailed estimates.

7. FITTINGS AND EQUIPMENT
(a) Fittings and Fixtures

Includes: All fittings and furnishings normally supplied under the construction contract.
Excludes: Equipment described in Element 7 (b)
Information is not usually available on fittings and fixtures at the time of preparing a preliminary estimate. If the cost of this element is recorded as a cost per square metre of the gross floor area whenever a detailed estimate is prepared, this cost can be used for preliminary estimates.

(b) Equipment

Includes: All equipment supplied and/or installed under the construction contract such as kitchen, laboratory, gymnasium, swimming pool, laundry, x-ray and sterilizing equipment, etc.
Excludes: Mechanical and electrical roughing in.
Obtain quotations from specialists who normally supply and/or install these items. Keep records of details and costs of special equipment. Costs can be recorded on a square metre basis for use in preliminary estimates as described for Element 7(a).

8. SERVICES
(a) Electrical

Includes: All the work inside the building normally included in the Electrical Division of the Specification.
Excludes: Electrical work outside the building.
Record and use the costs of the various electrical systems from detailed estimates as follows:
   i) Transformers and main switchboard: Lump sum costs to be used.

APPENDIX B

    ii) Lighting and power distribution: This includes distribution panels for light and power and all associated conduit, wiring and outlet boxes, and should be recorded as a cost per square metre of the gross floor area.
    (iii) Lighting fixtures: Enumerate light fixtures if possible, otherwise record the cost per square metre of the gross floor area.
    (iv) Branch wiring: Cost per square metre of the gross floor area.
    (v) Special services: This includes systems such as empty conduit for telephones, T.V. and radio systems, lightning conductors, fire alarms, under-floor ducts, clocks, P.A., emergency lighting equipment, stage lighting, doctors' and nurses' call, intercom, central dictating, automatic electric door locks and closers, car heaters, snow melting, telautograph, photofloods, etc. The extent of these special services must be known and the cost of each system estimated by obtaining quotations from suppliers or by reference to previous detailed estimates.

(b) Plumbing and Drains

Includes: All the plumbing and drainage work inside the building normally included in the Plumbing Division of the Specification.
Excludes: Outside drainage.
Record and use the costs of the plumbing systems from detailed estimates as follows:
    (i) Fixtures, Services and Wastes: This includes all sanitary fixtures, floor and roof drains, firehose cabinets, piping, hangers, insulation, fittings, valves, traps, etc. Enumerate plumbing fixtures if possible, otherwise record the cost per square metre of the gross floor area.
    (ii) Special Services: This includes systems such as liquid soap, compressed air, oxygen, vacuum, gas, etc. Record as lump sums or as the cost per square metre of the gross floor area for each system.
    (iii) Sprinklers: Cost per square metre of sprinkled area.

(c) Heating, Ventilation and Air Conditioning

Includes: All the heating, ventilation and air conditioning work normally included in the Mechanical Division of the Specification plus the cost of any heat storage tanks, fireplaces and chimneys. Record and use the costs of the mechanical systems from detailed estimates as follows:
    (i) Boiler plant: This includes boilers, boiler controls, fuel storage and supply, pumps, motors, etc. Record the cost per square

metre of the gross floor area, or per horsepower of the boilers.
(ii) **Refrigeration plant:** This includes the cooling tower, equipment, controls, motors, bases, anti-vibration pads, etc. Record the cost per square metre of the gross floor area, or per ton of refrigeration.
(iii) **Distribution systems:** This includes all supply and return piping, insulation, hangers, fittings, valves, radiators, convectors, etc. required to radiate or absorb heat. When possible enumerate radiators, convectors, etc., otherwise record the cost per square metre of the gross floor area.
(iv) **Ventilation system:** This includes all ductwork, fans, coils, dampers, grilles, sound and acoustic insulation, hangers, etc. Record the cost per square metre of the gross floor area being ventilated.
(v) **Controls:** Cost per square metre of the gross floor area.
(vi) **Chimney:** Cost per metre of height of the chimney.

9. SITE DEVELOPMENT
(a) General

Includes: Roads, paths, parking areas, curbs, bumpers, grading, seeding, sodding, planting and landscaping, retaining walls, canopies and covered ways [other than those included in Element 3(f)], pools, fencing and gates, permanent signs, outside steps, balustrades, paved terraces, etc.
Measure each type of area separately in square metres.
Measure curbs, bumpers, fences and balustrades in metres.

(b) Services

Includes: Incoming electrical services, connection to the electric main, water mains, septic tanks and tile beds, sewers, site drainage, manholes, catch basins and fountains.
Record the cost per square metre of the site for this work.

(c) Alterations

Includes: Any minor works of alteration to an existing building caused by erecting the new building next to it.
Excludes: Major alterations in an existing building.
Measure alterations in detail recording all structural removals in cubic metres, walls and partitions or finishes to be removed or replaced in square metres.
Enumerate and describe all openings to be cut or filled in, doors and

APPENDIX B

windows to be removed or installed, fittings and fixtures to be removed or replaced. Note that since the gross floor area is that of the new addition only, alterations are expressed as a cost per square metre of the new addition and not of the area of alterations.

(d) Demolition

Includes: All demolition of existing buildings to make way for new construction.

Demolition of existing buildings can only be done on a basis of known costs of similar jobs or by consulting a specialist in this type of work.

10. OVERHEAD AND PROFIT
(a) Site Overhead

Includes: The contractor's indirect and site expenses such as:
(i) Access to the Site — temporary roads and traffic control.
(ii) Site Accommodation — temporary offices, storage sheds, toilets, hoardings, signs, telephone, stationery and site office heating.
(iii) Site Protection — barricades, ladders, scaffolding, guard rails, fire precautions and safety precautions.
(iv) Temporary Services — water and hydro.
(v) Clean-Up — cleaning glass, cleaning site, cleaning building and removing rubbish.
(vi) Labour Expenses — overtime, lost and reporting time, strike delays, lodging and travel, and wage increases.
(vii) Supervision — superintendent, engineer, watchman and other site staff.
(viii) Insurance and Bonds — fire insurance, public liability and property damage insurance, performance bond, and bid bond.
(ix) Equipment — elevator and hoist, crane, trucks, buggies and vibrators, pumps, fuel and oil, and small tools.
(x) Winter Conditions — winter heat, winter concrete, premium for winter work, clearing ice and snow, tarpaulins, straw, and enclosures etc.
(xi) Miscellaneous — building permit, land survey, layout and preparation, attendance on architect, attendance on subcontractors, critical path, rough hardware, testing and samples, photographs, cutting and patching, and financing.

For preliminary estimates assess this as a percentage of the direct costs, that is, the work in Elements 1(a) - 9(d) inclusive.

(b) Head Office Overhead and Profit

11. CONTINGENCIES

(a) Design Contingency

The design contingency is intended to provide an allowance so that necessary design changes can be made as the design is developed. It may therefore be as high as ten percent of the estimated construction cost — Elements 1(a) - 10 (b) inclusive — in the preliminary stages, reducing to zero for the final estimate before tenders are called.

(b) Escalation Contingency

The escalation contingency is intended to allow for increases in costs between the date the estimate is prepared and the date tenders are called. (Note: The escalation normally carried by the contractor in his tender for increases in costs between the date of his tender and the time the work is carried out should be included in the estimate and should not be allowed for here.)

(c) Post-Contract Contingency

The post-contract contingency is intended to provide an allowance so that changes can be made during construction. The amount of this contingency is left to the discretion of the architect and his client.

APPENDIX C

 Helyar & Associates
Helyar, Rae, Mauchan & Hall Limited
Chartered Quantity Surveyors · Construction Consultants

OUTLINE SPECIFICATION

PROJECT _____

LOCATION _____

BUILDING TYPE _____

ARCHITECT _____

    Representative _____ Tel. No. _____

    Architect's Job No. _____

MECHANICAL ENGINEER _____

    Representative _____ Tel. No. _____

ELECTRICAL ENGINEER _____

    Representative _____ Tel. No. _____

STRUCTURAL ENGINEER _____

    Representative _____ Tel. No. _____

OTHER CONSULTANT _____

    Representative _____ Tel. No. _____

H & A JOB NO. _____

ESTIMATE NO. _____   DATE _____

Please return to HELYAR & ASSOCIATES

207 Avenue Road
Toronto, Ontario
M5R 2J3
416-961-2261

Suite 120
800 Windmill Rd.,
Dartmouth, N. S.
B3B 1L1
(902) 466-2702

Suite 114
370 Metcalfe Street
Ottawa, Ontario
K2P 1S9
613-238-5023

CONSTRUCTION ESTIMATING AND COSTING

This document is intended primarily as a checklist pre-tender estimate, although it can also be used as an outline specification to accompany a detailed cost analysis. When used for the former purpose it is recognized that it will not always be possible to complete it fully, but as much information as is available should be given, and where assumptions have been made they should be so noted.

GENERAL INFORMATION

(1) ESTIMATING SCHEDULE - This estimate due:_____ 19_____

                      Anticipated Tender Date:_____ 19_____

(2) ANTICIPATED DURATION OF CONSTRUCTION:_____ Months

(3) TYPE OF CONTRACT:_____

(4) SALES TAX - Provincial Sales Tax: To be included ( )   Exempt ( )

                Federal Sales Tax: To be included ( )   Exempt ( )

- 2 -

APPENDIX C

(1) SUBSTRUCTURE

    (a) NORMAL FOUNDATIONS

        (1) Soil Conditions: _____

        (2) Type of Foundation: _____

        (3) Disposal of Excavated Material: _____

        (4) Type of Backfill: _____

        (5) Foundation Walls: _____

        (6) Concrete Strength: Wall Footings _____

                                      Column Footings _____

                                      Walls _____

                                      Columns _____

        (7) Weeping Tile: _____

        (8) Comments: _____

            _____

            _____

            _____

    (b) BASEMENT

        (1) Average Depth Below Existing Ground Level: _____

        (2) Type of Backfill: _____

        (3) Comments: _____

            _____

            _____

            _____

CONSTRUCTION ESTIMATING AND COSTING

1. SUBSTRUCTURE (CONTINUED)

   (c) SPECIAL FOUNDATIONS

      (1) Rock Excavation: _____

      (2) Water Conditions: _____

      (3) Sheet Piling: _____

      (4) Piles: _____

      (5) Caissons: _____

      (6) Underpinning: _____

      (7) Comments: _____
_____
_____
_____

2. STRUCTURE

   (a) LOWEST FLOOR CONSTRUCTION

      (1) Type of Slab: _____

      (2) Concrete Slab: Thickness: _____ Concrete Strength: _____

      (3) Fill Under Slab: _____

      (4) Vapour Barrier: _____

      (5) Waterproofing: _____

      (6) Skim Coat: Thickness: _____ Concrete Strength: _____

      (7) Reinforcing: Type: _____

                        Mass: _____ kg per m^2

      (8) Comments: _____
_____
_____
_____

- 4 -

APPENDIX C

2. STRUCTURE (CONTINUED)

   (b) UPPER FLOOR CONSTRUCTION

   (1) Type of Construction: _____
   _____
   _____

   (2) Average Bay Size or Span: _____

   (3) Superimposed Load: _____

   (4) Average Slab Thickness: _____

   (5) Average Column Size: _____

   (6) Concrete Strength: Slabs and Beams: _____

                                   Columns: _____

   (7) Steel Framing: Structural Steel: _____ kg per m^2

                              Steel Joists: _____ kg mer m^2

   (8) Reinforcing Steel: Beams and slabs only: _____ kg per m^2

                                  Columns only: _____ kg per m

                                  Beams, slabs and columns: _____ kg per m^2

   (9) Metal Deck: _____

   (10) Fireproofing: Slabs: _____

                              Beams: _____

                              Columns: _____

   (11) Catwalks: _____

   (12) Comments: _____
   _____
   _____
   _____

CONSTRUCTION ESTIMATING AND COSTING

2. STRUCTURE (CONTINUED)

   (c) ROOF CONSTRUCTION

   (1) Type of Construction: _____
   _____
   _____

   (2) Average Bay Size or Span: _____

   (3) Superimposed Load: _____

   (4) Average Slab or Deck Thickness: _____

   (5) Average Column Size: _____

   (6) Concrete Strength: Slabs and Beams: _____

                                   Columns: _____

   (7) Steel Framing: Structural Steel: _____ kg per m^2

                            Steel Joists: _____ kg per m^2

   (8) Reinforcing Steel: Beams and slabs only: _____ kg per m^2

                            Columns only: _____ kg per m

                            Beams, slabs and columns: _____ kg per m^2

   (9) Metal Deck: _____

   (10) Fireproofing: Slab or Deck: _____

                         Beams: _____

                         Columns: _____

   (11) Comments: _____
   _____
   _____
   _____

- 6 -

APPENDIX C

3. EXTERIOR CLADDING

  (a) ROOF FINISH

    (1) Type of Roof: _____

    (2) Material: _____

    (3) Insulation: _____

    (4) Vapour Barrier: _____

    (5) Flashings: _____

    (6) Projections: Roof Hatches: _____

                          Roof Domes: _____

                          Skylights: _____

                          Duckboards: _____

    (7) Parapet Walls: Material: _____

                          Height: _____

    (8) Comments: _____
                 _____
                 _____

  (b) WALLS BELOW GROUND FLOOR

    (1) Type of Wall: _____
        _____ Thickness: _____

    (2) Concrete Strength: _____

    (3) Reinforcing: _____

    (4) Waterproofing: _____

    (5) Dampproofing: _____

    (6) Comments: _____
                 _____
                 _____
                 _____

CONSTRUCTION ESTIMATING AND COSTING

3. <u>EXTERIOR CLADDING</u> (CONTINUED)

   (c) WALLS ABOVE GROUND FLOOR

      (1) Type of Wall: _____

           _____ Thickness: _____

      (2) Facing or Exterior Finish: _____

      (3) Back-up: _____

      (4) Insulation: _____

      (5) Vapour Barrier: _____

      (6) Comments: _____

              _____

              _____

              _____

   (d) WINDOWS

      (1) Type: _____

      (2) Material: _____

      (3) Finish: _____

      (4) Glass: _____

      (5) Blinds or Drapes: _____

      (6) Louvres: _____

      (7) Hardware: _____

      (8) Comments: _____

              _____

              _____

              _____

- 8 -

APPENDIX C

3. <u>EXTERIOR CLADDING</u> (CONTINUED)

   (e) EXTERIOR DOORS AND SCREENS

      (1) Exterior Doors: _____

           Frames: _____

      (2) Exterior Screens: _____

           Finish: _____

           Glass: _____

           Screen Doors: _____

      (3) Hardware: _____

      (4) Comments: _____

           _____

           _____

           _____

   (f) BALCONIES AND PROJECTIONS

      (1) Roof Projections: _____

      (2) Floor Projections: _____

      (3) Balconies: _____

      (4) Canopies: _____

      (5) Railings: _____

      (6) Comments: _____

           _____

           _____

           _____

CONSTRUCTION ESTIMATING AND COSTING

4. INTERIOR VERTICAL ELEMENTS

   (a) PERMANENT PARTITIONS AND DOORS

      (1) Partitions: _____

      _____

      _____

      (2) Toilet Partitions: _____

      (3) Doors: _____

      _____

           Frames: _____

           Door Finish: _____

      (4) Hardware: _____

      (5) Comments: _____

      _____

      _____

      _____

   (b) MOVABLE PARTITIONS AND DOORS

      (1) Partitions: _____

      _____

      (2) Doors: _____

      (3) Comments: _____

      _____

      _____

      _____

APPENDIX C

4. INTERIOR VERTICAL ELEMENTS (CONTINUED)

   (c) GLAZED PARTITIONS AND DOORS

      (1) Partitions: _____

_____

_____

      (2) Doors: _____

      (3) Comments: _____

_____

_____

_____

5. VERTICAL MOVEMENT

   (a) STAIRS

      (1) Material: _____

      (2) Finish: _____

      (3) Railings: _____

            Handrails: _____

      (4) Steps: _____

      (5) Ladders: _____

      (6) Comments: _____

_____

_____

_____

CONSTRUCTION ESTIMATING AND COSTING

5. <u>VERTICAL MOVEMENT</u> (CONTINUED)

   (b) ELEVATORS AND ESCALATORS

      (1) Passenger Elevators: Capacity: _____ Speed: _____

                                     Type: _____

      (2) Service Elevators: Capacity: _____ Speed: _____

                                       Type: _____

      (3) Dumbwaiters: Capacity: _____ Speed: _____

                                       Type: _____

      (4) Hoists: Capacity: _____ Speed: _____

                                       Type: _____

      (5) Escalators: Capacity: _____ Width: _____

                                       Speed: _____

      (6) Comments: _____

                _____

                _____

                _____

APPENDIX C

6. INTERIOR FINISHES

   (a) FLOOR FINISHES

      (1) Floors

	%
Concrete: Steel Trowel	____ %
Hardener	____ %
Painted	____ %
Vinyl Asbestos Tile	____ %
High Content Vinyl Tile	____ %
Sheet Vinyl	____ %
Rubber Tile	____ %
Sheet Rubber	____ %
Quarry Tile	____ %
Ceramic Tile	____ %
Terrazzo	____ %
Epoxy	____ %
Stone Paving	____ %
Slate Paving	____ %
Marble	____ %
Wood	____ %
Parquet	____ %
Carpet	____ %
_____	____ %
	100 %

Topping _____ Thick ____ %

      (2) Base

	%
Cement	____ %
Rubber	____ %
Vinyl	____ %
Quarry Tile	____ %
Ceramic Tile	____ %
Terrazzo	____ %
Stone	____ %
Slate	____ %
Marble	____ %
Wood	____ %
Metal	____ %
_____	____ %
	100 %

      (3) Comments: _____

_____

_____

_____

- 13 -

CONSTRUCTION ESTIMATING AND COSTING

6. INTERIOR FINISHES (CONTINUED)

   (b) CEILING FINISHES

   (1) Ceilings

Exposed Structure	_____ %	of which: _____	% painted
Painted Plaster	_____ %	of which: _____	% suspended
Painted Plasterboard	_____ %	of which: _____	% suspended
Acoustic Tile	_____ %	of which: _____	% suspended
Wood	_____ %	Type:	
Metal	_____ %	Type:	
Illuminated	_____ %	Type:	
Bulkheads	_____ %	Type:	
_____	_____ %		
	100 %		

   (2) Cornices: _____

   (3) Comments: _____
   _____
   _____
   _____

   (c) WALL FINISHES

   (1) Finishes

Exposed Block or concrete	_____ %	of which: _____	% painted
Plaster	_____ %	of which: _____	% painted
			_____ % wall covering
Plasterboard	_____ %	of which: _____	% painted
			_____ % wall covering
Acoustic Plaster	_____ %		
Ceramic Tile	_____ %		
Wood Panelling	_____ %		
Brick Facing	_____ %		
Marble Facing	_____ %		
_____	_____ %		
	100 %		

   (2) Corner Guards: _____
   (3) Dado Rails: _____
   (4) Comments: _____
   _____
   _____
   _____

APPENDIX C

7. FITTINGS AND EQUIPMENT

    (a) **FIT**TINGS AND FIXTURES

    Locate on drawings and identify by number:

(1)	Ash Trays	( )	(21)	Musical Instrument Racks	( )
(2)	Benches	( )	(22)	Notice Boards	( )
(3)	Bleachers	( )	(23)	Nurses' Stations	( )
(4)	Carrels	( )	(24)	Parking Booths	( )
(5)	Chalkboards	( )	(25)	Seats	( )
(6)	Coat Racks	( )	(26)	Shelving, Wood	( )
(7)	Counters	( )	(27)	Shelving, Metal	( )
(8)	Cupboards, Wood	( )	(28)	Tackbo**ards**	( )
(9)	Cupboards, Metal	( )	(29)	Washroom Accessories	( )
(10)	Curtain Track	( )	(30)	Waste Receptacles	( )
(11)	Directory Boards	( )	(31)	_____	( )
(12)	Information Booths	( )	(32)	_____	( )
(13)	Janitors' Shelves	( )	(33)	_____	( )
(14)	Kitchen Counters	( )	(34)	_____	( )
(15)	Kitchen Cupboards	( )	(35)	_____	( )
(16)	Library Shelving	( )	(36)	_____	( )
(17)	Lockers	( )	(37)	_____	( )
(18)	Mail Boxes	( )	(38)	_____	( )
(19)	Milk Cabinets	( )	(39)	_____	( )
(20)	Mirrors	( )	(40)	_____	( )

    (41) Comments: _____
    _____
    _____
    _____

CONSTRUCTION ESTIMATING AND COSTING

7. FITTINGS AND EQUIPMENT (CONTINUED)

   (b) Equipment:

Autopsy Tables:	In contract ( )	NIC ( )	None ( )
Bed Pan Racks:	" " ( )	" ( )	" ( )
Bowling Alley Equipment:	" " ( )	" ( )	" ( )
Cage Washing Equipment:	" " ( )	" ( )	" ( )
Chart Racks:	" " ( )	" ( )	" ( )
Conveyors:	" " ( )	" ( )	" ( )
Cranes and Monorails:	" " ( )	" ( )	" ( )
Cyclarama:	" " ( )	" ( )	" ( )
Dock Levellers:	" " ( )	" ( )	" ( )
Dryers:	" " ( )	" ( )	" ( )
Fireproof Curtains:	" " ( )	" ( )	" ( )
Garbage Disposal:	" " ( )	" ( )	" ( )
Glassware Washing:	" " ( )	" ( )	" ( )
Greenhouse:	" " ( )	" ( )	" ( )
Gymnasium Equipment:	" " ( )	" ( )	" ( )
Hydro-Therapy Tanks:	" " ( )	" ( )	" ( )
Incinerators:	" " ( )	" ( )	" ( )
Kitchen Equipment:	" " ( )	" ( )	" ( )
Laboratory Equipment:	" " ( )	" ( )	" ( )
Laundry Equipment:	" " ( )	" ( )	" ( )
Medical Refrigerators:	" " ( )	" ( )	" ( )
Mortuary Refrigerators:	" " ( )	" ( )	" ( )
Operating Tables:	" " ( )	" ( )	" ( )
Pathological Incinerators:	" " ( )	" ( )	" ( )
Pneumatic Tube System:	" " ( )	" ( )	" ( )
Projection Room Equipment:	" " ( )	" ( )	" ( )
Projection Screen:	" " ( )	" ( )	" ( )
Refrigerators:	" " ( )	" ( )	" ( )
Sauna Equipment:	" " ( )	" ( )	" ( )
Scenery Hangers:	" " ( )	" ( )	" ( )
School Shop Equipment:	" " ( )	" ( )	" ( )
Solution Cabinets:	" " ( )	" ( )	" ( )
Sonic Energy Equipment:	" " ( )	" ( )	" ( )
Stage Equipment:	" " ( )	" ( )	" ( )
Sterilizers:	" " ( )	" ( )	" ( )
Stills:	" " ( )	" ( )	" ( )
Stoves:	" " ( )	" ( )	" ( )
Swimming Pool Equipment:	" " ( )	" ( )	" ( )
Truck Washing Equipment:	" " ( )	" ( )	" ( )
Vending Machines:	" " ( )	" ( )	" ( )
Viewing Boxes:	" " ( )	" ( )	" ( )
Washing Machines:	" " ( )	" ( )	" ( )
Window Washing Equipment:	" " ( )	" ( )	" ( )
X-Ray Equipment:	" " ( )	" ( )	" ( )
_____	" " ( )	" ( )	" ( )
_____	" " ( )	" ( )	" ( )
_____	" " ( )	" ( )	" ( )

   (2) Comments: _____

_____

_____

- 16 -

APPENDIX C

8. SERVICES

    See Separate Outline Specification.

9. SITE DEVELOPMENT

    (a) GENERAL

        (1) Area of the Site: _____
        (2) Topography: _____
        (3) Clearing and Grubbing: _____
        (4) Grading: _____
        (5) Paving: _____
        (6) Curbs: _____
        (7) Bumpers: _____
        (8) Parking Lines: _____
        (9) Steps: _____
       (10) Seeding: _____
       (11) Sodding: _____
       (12) Planting: _____
       (13) Pools: _____
       (14) Fountains: _____
       (15) Tennis Courts: _____
       (16) Running Track: _____
       (17) Playground: _____
       (18) Permanent Signs: _____
       (19) Retaining Walls: _____
       (20) Fences: _____
       (21) Flagpoles: _____
       (22) Canopies and Covered Ways: _____

       (23) Landscaping Allowance: $ _____
       (24) Other: _____
       (25) Comments: _____

_____
_____
_____

- 17 -

CONSTRUCTION ESTIMATING AND COSTING

9. <u>SITE DEVELOPMENT</u> (CONTINUED)

   (b) SERVICES

      (1) Electrical: _____

      (2) Water: _____

      (3) Drains & Sewers: _____

      (4) Comments: _____

      _____

      _____

      _____

   (c) ALTERATIONS

      (1) Alterations in the existing building:

            In contract ( )   NIC ( )   None ( )

      (2) Comments: _____

      _____

      _____

      _____

   (d) DEMOLITION

      (1) Existing Buildings: To Remain   ( )

                                    To be Demolished   ( )

                                    In Contract ( ) NIC ( )

      (2) Comments: _____

      _____

      _____

      _____

APPENDIX C

10. OVERHEAD AND PROFIT

    (1) Performance Bond: _____ %

    (2) Bid Bond: _____

    (3) Payment Bond: _____

    (4) Fire Insurance: By Owner ( )   By Contractor ( )

    (5) Building Permit: By Owner ( )   By Contractor ( )

    (6) Critical Path: _____

    (7) Photographs: _____

    (8) Comments: _____

                       _____

                       _____

                       _____

11. CONTINGENCIES

    (a) Design Contingency   Allow _____ %

    (b) Escalation Contingency   Allow _____ %

    (c) Post-Contract Contingency Allow _____ %

CONSTRUCTION ESTIMATING AND COSTING

CLARE, RANDALL-SMITH & ASSOCIATES LIMITED

53 LESMILL ROAD, DON MILLS, ONTARIO
(416) 445-8166

ELEMENT SERVICES

OUTLINE SPECIFICATION

PROJECT   _____
PROJECT NO.   _____
DATE   _____
CLIENT   _____
  REPRESENTATIVE   _____ Tel. No. _____
LOCATION   _____
BUILDING TYPE   _____
ARCHITECT   _____
  REPRESENTATIVE   _____ Tel. No. _____
MECHANICAL ENGINEER   _____
  REPRESENTATIVE   _____ Tel. No. _____
ELECTRICAL ENGINEER   _____
  REPRESENTATIVE   _____ Tel. No. _____
STRUCTURAL ENGINEER   _____
  REPRESENTATIVE   _____ Tel. No. _____

CONSTRUCTION PERIOD   _____
COMMENCEMENT DATE   _____
NUMBER OF FLOORS   _____
GROSS FLOOR AREA   _____
INDOOR PARKING AREA   _____

NO. OF HOSPITAL BEDS   _____

MECHANICAL & ELECTRICAL QUANTITY SURVEYORS AND CONSTRUCTION ECONOMISTS

- 1 -

APPENDIX C

8. SERVICES

   (a) ELECTRICAL

     (1) H.V. SWITCHBOARD

          New_____ Existing_____ Revised_____

          Fused or Circuit Breaker_____

          Size of Fuses or C.B.'s_____

          Request sketch showing make-up of cubicles

     (2) TRANSFORMERS

          Size_____KVA

          Type (dry, oil or askarel filled)_____

     (3) HIGH VOLTAGE DISTRIBUTION FEEDERS

          Voltage_____ Type_____

          Feeding_____

     (4) LOW VOLTAGE DISTRIBUTION FEEDERS

          Voltage_____ Type_____

          Feeding_____

     (5) L.V. SWITCHBOARDS

          New_____ Existing_____ Revised_____

          Voltage_____Volts Mains Current_____Amps

          Circuit Breakers_____ Type_____

          Fused_____

          Main Switch or CB Size_____

          No. of Sections_____

          Interrupting Capacity_____

CLARE. RANDALL-SMITH & ASSOCIATES LIMITED

- 2 -

CONSTRUCTION ESTIMATING AND COSTING

8. SERVICES (cont'd)

   (a) ELECTRICAL (cont'd)

      (6) UNIT SUB STATION

          Type of Incoming Line Section_____

          Type of Transformer_____

          Type of Outgoing Section_____

          Indoor_____

          Outdoor_____

          Interrupting Capacity_____

      (7) STANDBY GENERATOR

          New_____ Existing_____

          Size_____KW Voltage_____Volts

          Manufacturer_____

          Fuel_____

          Transfer Switch_____

          Exhaust System_____

      (8) BUS DUCT

          Copper or Aluminum_____ Type_____

          Capacity_____ Amps

          Poles_____ Neutral_____

          Feeding_____

CLARE, RANDALL-SMITH & ASSOCIATES LIMITED

- 3 -

APPENDIX C

8. SERVICES (cont'd)

   (a) ELECTRICAL (cont'd)

   (9) MOTOR CONTROLLERS

        MCC_____ Sep. Starters_____

        MCC Class_____

        No. of Sections_____

        Fused or Breakers_____

(10) MOTOR WIRING

        No. of Motors_____ Voltage_____Volts

        Total H.P._____

        Type of Air Conditioning_____

        A/C Voltage_____

(11) SECONDARY FEEDERS

        Voltage_____ Type_____

        Request riser diagram

(12) POWER PANELS

        Voltage_____ Capacity_____

        Circuit Breaker_____ Fused_____

(13) LIGHTING PANELS

        Voltage_____ Type_____

CLARE, RANDALL-SMITH & ASSOCIATES LIMITED

CONSTRUCTION ESTIMATING AND COSTING

8. SERVICES (cont'd)

   (a) ELECTRICAL (cont'd)

      (14) FIXTURES

          Voltage_____Type of Lamps_____

          Main Fixture_____

          Type of Lens_____

          LV Control_____Switching (local or panel)_____

          Dimming_____

          Morality_____

          Special_____

          Auditorium Lighting & Dimming:

                  Control Console_____

                  Amplifier Bank_____

                  Patch Panel_____

                  Remote Control Panels_____

                  Preset Banks_____

                  Battens_____

                  Type of Lights_____

      (15) BRANCH WIRING

          Wiring Type_____

CLARE, RANDALL-SMITH & ASSOCIATES LIMITED

APPENDIX C

8. SERVICES (cont'd)

   (a) ELECTRICAL (cont'd)

   (16) UNDERFLOOR DUCT

   Manufacturer_____

   Trench Duct_____ Width_____

   Header Duct_____

   Flush Duct_____ No. of Ducts_____ Sizes_____

   Insert Duct_____ No. of Ducts_____ Sizes_____

   Fittings by Contractor_____ By Tenant_____

   Duct Spacing_____

   (17) ELECTRIC HEATING

   Type_____

   Voltage_____

   Control_____

   Peak - Load Control_____

   (18) FIRE ALARM

   New_____ Existing_____

   Manufacturer_____

   Type_____

   Conduit Type_____

   Smoke Detection_____ Ceiling_____ Ducts_____

   AC or DC_____ Zoned_____

   Coded_____ Pre-Signal_____

   P.A. Facility_____

   Indoor Parking_____

CLARE, RANDALL - SMITH & ASSOCIATES LIMITED

CONSTRUCTION ESTIMATING AND COSTING

8. SERVICES (cont'd)

   (a) ELECTRICAL (cont'd)

   (19) CLOCK & PROGRAM

   New_____ Existing_____

   Manufacturer_____

   Type_____

   Program_____

   (20) PA SYSTEM

   New_____ Existing_____

   Manufacturer_____

   Type_____

   Areas Services_____

   Auditorium System_____

   (21) EMERGENCY LIGHTING

   From Standby Generator_____

   Battery Units_____

   Central Battery Unit_____

   (22) LIGHTNING PROTECTION

   Type_____

   (23) TELEPHONE

   Type of Service_____

   Distribution - Cable Tray_____ Size_____

   Floor(s) Affected_____

   Conduit_____ Type_____

CLARE, RANDALL-SMITH & ASSOCIATES LIMITED

-7-

APPENDIX C

8. SERVICES (cont'd)

   (a) ELECTRICAL (cont'd)

      (24) INTERNAL TELEPHONE (PAX)

           New_____ Existing_____

           Manufacturer_____

           No. of Lines_____

           No. of Telephones_____

      (25) INTERCOMM SYSTEMS

           New_____ Existing_____

           Manufacturer_____

           No. of Stations_____

           No. of Lines on Control Station_____

      (26) ENTERTAINMENT TV

           New_____ Existing_____

           Empty Conduit System_____

           Antenna & Amplifiers_____

           Wiring_____

           Cable TV_____

CLARE, RANDALL-SMITH & ASSOCIATES LIMITED
-8-

CONSTRUCTION ESTIMATING AND COSTING

8. SERVICES (cont'd)

   (a) ELECTRICAL (cont'd)

   (27) CLOSED CIRCUIT TV

   New_____ Existing_____

   Empty Conduits Only_____

   Equipment_____

   Manufacturer_____

   Locations Served_____

   Teaching_____

   Monitoring_____

   Studio_____

   Surveillance_____

   Special Lighting_____

   (28) ENTERTAINMENT RADIO

   New_____ Existing_____

   Empty Conduit System_____

   Radio_____

   Muzak_____

   (29) KITCHEN WIRING

   New_____ Existing_____

   Area_____

   No. of People Served_____

CLARE, RANDALL-SMITH & ASSOCIATES LIMITED
-9-

APPENDIX C

8. SERVICES (cont'd)

   (a) ELECTRICAL (cont'd)

      (30) LABORATORY WIRING

           New_____ Existing_____

           Type_____

           Area_____

      (31) NURSE CALL

           New_____ Existing_____

           Manufacturer_____

           Type_____

      (32) BED AVAILABILITY

           New_____ Existing_____

           Manufacturer_____

           Riser_____

      (33) DOCTORS REGISTER

           New_____ Existing_____

           Manufacturer_____

           Type_____

           No. of Locations_____

           Operators Console_____

      (34) RADIO PAGING

           New_____ Existing_____

           Call Capacity_____

CLARE, RANDALL-SMITH & ASSOCIATES LIMITED

CONSTRUCTION ESTIMATING AND COSTING

8. SERVICES (cont'd)

   (a) ELECTRICAL (cont'd)

   (35) DICTATION SYSTEM

       New_____ Existing_____

       No. of Locations_____

       Empty Conuits Only_____

       Equipment_____

       Manufacturer_____

   (36) CART SYSTEMS

       Type_____

   (37) DOCTOR PAGING

       New_____ Existing_____

       Manufacturer_____

       Type_____

   (38) MONITORING SYSTEMS

       Physiological Monitor_____
       (heart, temperature, respiration, blood)

   (39) ALARMS

       Oxygen_____

       Nitrous Oxide_____

       Drug & Narcotics_____

       Blood Bank_____

       Nurse Supervisory Door Alarm_____

       Nurses Residence_____

CLARE, RANDALL-SMITH & ASSOCIATES LIMITED
-11-

APPENDIX C

8. SERVICES (cont'd)

   (a) ELECTRICAL (cont'd)

      (40) X-RAY ROOMS

           Number _____

      (41) OPERATING ROOMS

           Number _____

           Ground Detection System _____

      (42) MISCELLANEOUS SYSTEMS

           Watchman's Tour _____

           Burglar Alarm _____

           Ground Fault Protection (Distribution) _____

           Computer Room _____

           R.F. Rooms _____

           Cable Tray _____ System _____

           Janitors Call System _____

           Hand Dryers in Washrooms _____

           Refrigerated Garbage _____

           Incinerator _____

           Elapsed Time Clock _____

      (43) GENERAL

           Conduit Type _____

           Wire Type _____

           Services/systems Connections to other Buildings ____

           Contingencies _____

CLARE, RANDALL-SMITH & ASSOCIATES LIMITED
-12-

CONSTRUCTION ESTIMATING AND COSTING

8. SERVICES (cont'd)

   (a) ELECTRICAL (cont'd)

     (44) SITE DEVELOPMENT

        Service:

           New_____ Existing_____ Revised_____

           Voltage_____ Volts

           Capacity_____ Amps

           Interrupting Capacity_____

           Cable Size_____ Type_____

           Overhead_____

           Underground_____ No. & Type of Ducts_____

           Distance to Hydro Source_____

           Manholes_____

           Service Tunnel_____

           What will the hydro supply?_____

           Request single line diagram

        Area Lighting:

           Street_____ Type_____ Length_____

           Car Park_____ Type_____ Area_____

           Walkways_____ Type_____ Length_____

           Architectural_____ Type_____

           Voltage_____

           Type of Poles_____

CLARE, RANDALL-SMITH & ASSOCIATES LIMITED

APPENDIX C

8. SERVICES (cont'd)

    (a) ELECTRICAL (cont'd)

    (44) SITE DEVELOPMENT (cont'd)

        Snow Melting:

            Location_____Area_____

            Voltage_____Volts

            Type of Control_____

    (45) COMMENTS   _____

_____

_____

_____

_____

_____

_____

_____

CLARE, RANDALL - SMITH & ASSOCIATES LIMITED

CONSTRUCTION ESTIMATING AND COSTING

8. SERVICES (cont'd)

   (b) PLUMBING AND DRAINS

   ---

   (1) INSIDE BURIED PIPING

       Storm   _____

       Sanitary & Vent   _____

       Water   _____

       Drainage products   _____

       Sub soil drain (weepers)   _____

   (2) ABOVE GROUND PIPING-STANDARD

       Rain water leaders and hoppers   _____

       Sanitary waste & vent   _____

       Drainage products   _____

       Shok-stop or air cushion   _____

       Domestic cold water   _____

       Domestic hot water   _____

       Domestic hot water recirculating   _____

       Fire Stand pipe system   _____

       Other   _____

   _____

   _____

   _____

   _____

   _____

   _____

CLARE, RANDALL-SMITH & ASSOCIATES LIMITED

APPENDIX C

8. SERVICES (cont'd)

   (b) PLUMBING AND DRAINS (cont'd)

---

      (3) ABOVE GROUND PIPING–SPECIAL

         Chilled drinking water   _____

         High temperature water 180°   _____

         Tempered water 110°   _____

         Sprinkler piping (wet or dry)   _____

         Sprinkler siporex   _____

         CO^2 System   _____

         Oxygen   _____

         Vacuum   _____

         Medical Air   _____

         Nitrous oxide   _____

         Nitrogen   _____

         De-ionized water   _____

         De-mineralized water   _____

         Compressed air   _____

         Distilled water   _____

         Laboratory waste & vent   _____

         Formlin piping   _____

         Gas piping   _____

         Garbage disposal system   _____

         Kitchen equipment piping   _____

         Other   _____

                              _____

                              _____

                              _____

CLARE, RANDALL-SMITH & ASSOCIATES LIMITED

CONSTRUCTION ESTIMATING AND COSTING

8. SERVICE (cont'd)

   (b) PLUMBING AND DRAINS (cont'd)

      (4) REGULAR FIXTURES

         W.C. (Tank) (Floor) (Wall)   _____

         W.C. (F.V.) (Floor) (Wall)   _____

         W.C. Carrier   _____

         Bidet   _____

         Urinal (Stall) (Wall) (PV) (Tank)   _____

         Urinal Carrier   _____

         Lavatory (W.H.)   _____

         Lavatory Carrier   _____

         Lavatory (C.I.)   _____

         Bradley Washfountains   _____

         Bath   _____

         Shower (Safe)   _____

         Drinking Fountain   _____

         Sink   _____

         Slop Sink   _____

         Mop Sink (Safe)   _____

         Laundry Tub   _____

         Washing Machine   _____

         Dishwasher   _____

         Washroom Accessories   _____

         Soap Systems   _____

         Garbage Can Washer   _____

         Other   _____

                                _____

CLARE, RANDALL-SMITH & ASSOCIATES LIMITED

APPENDIX C

8. SERVICES (cont'd)

   (b) PLUMBING AND DRAINS (cont'd)

      (5) HOSPITAL FIXTURES

          Bed Pan Cleanser _____

          Bed Pan Sterilizer _____

          Bath - Institutional _____

          Bath - Emergency _____

          Bath - Arm _____

          Bath - Sitz _____

          Bath - Foot _____

          Bath - Pre-Natal _____

          Bath - Infants _____

          Bath - Continuous flow _____

          Bath - Hydro Therapy _____

          Bath - Underwater Treatment _____

          Therapy Swim Pool _____

          Sink - Scrub up _____

          Sink - Plaster & Trap _____

          Sink - Flushing rim _____

          Sink - Medicine _____

          Sink - Utility _____

          Autoclave _____

          Sterilizer _____

          Autopsy Table _____

          Flushing Rim Floor Drains _____

          Laboratory Fitments _____

          Other Sink Carriers _____

CLARE, RANDALL-SMITH & ASSOCIATES LIMITED

CONSTRUCTION ESTIMATING AND COSTING

8. SERVICES (cont'd)

   (b) <u>PLUMBING AND DRAINS (cont'd)</u>

      (6) <u>EQUIPMENT</u>

Sewage Pumps	_____
Sump pumps	_____
Sumbersible pump	_____
Domestic hot water tanks	_____
Domestic hot water heater	_____
Domestic re-circ. pumps	_____
Domestic cold water booster pumps	_____
Domestic hot water booster pumps	_____
Domestic water controls	_____
Phneumatic booster tank	_____
Pressure reducing valves	_____
Chiller for drinking water	_____
Fire Pump	_____
Fire pump controls	_____
Fire hose cabinets	_____
Fire blankets & F.B. cabinet	_____
Fire extinguisher wall hung	_____
Fire extinguisher w/F.E. cabinets	_____
Fire storage tanks	_____
Siamese connections	_____
Drainage products F.D.	_____
Roof drains	_____
Trap seals	_____
Outside H.B.	_____
Oil interceptors	_____
Grease interceptors	_____
Somat garbage unit	_____
Hydro guards mix. valves	_____
Water Stills	_____
Water Treatment (Softeners etc)	_____

CLARE, RANDALL-SMITH & ASSOCIATES LIMITED

APPENDIX C

8. SERVICES (cont'd)

   (b) PLUMBING AND DRAINS (cont'd)

      (7) SUB-CONTRACTS

          Sprinklers (wet or dry) _____

          Sprinklers (Siporex) _____

          Isolation _____

          Insulation _____

          Kitchen refrigeration _____

          Incinerator (waste) _____

          Incinerator (Pathological) _____

          Concrete work (bases etc) _____

          Cover & frames _____

          Steel gratings _____

          Starters & wiring _____

          Painting _____

          Hoisting _____

          Vacuum cleaning system _____

          Phneumatic tube system _____

          Kitchen Equipment _____

          Laundry Equipment _____

CLARE, RANDALL-SMITH & ASSOCIATES LIMITED

CONSTRUCTION ESTIMATING AND COSTING

8. SERVICES (cont'd)

   (b) <u>PLUMBING AND DRAINS (cont'd)</u>

      (8) <u>MISCELLANEOUS</u>

Permits	_____
Equipment Drains	_____
Testing & Clean-up	_____
Incidentals	_____
Sleeves	_____
Tags & Directory	_____
Cutting & Patching	_____
Misc. Steel	_____
Access Doors	_____
Cash Allowances	_____
Bonds	_____
Kitchen Hook-ups	_____
Laboratory hook-ups	_____
Freight or Cartage	_____
Temporary Water	_____
Temporary W.C.'s	_____
Temporary Fire Protection	_____
Contingencies	_____

CLARE, RANDALL-SMITH & ASSOCIATES LIMITED

APPENDIX C

8. SERVICES (cont'd)

   (b) PLUMBING AND DRAINS (cont'd)

      (9) SITE DEVELOPMENT

         Service Connections:

            Storm   _____

            Outfall   _____

            Sanitary   _____

            Septic Tank & Bed   _____

            Domestic Water   _____

            Fire   _____

            Gas   _____

            Others   _____

            Excavation & Backfill   _____

         Outside Services:

            Storm (w/manholes & catch
                  basins)   _____

            Parking lot drainage   _____

            Sanitary (w/manholes)   _____

            Water Service & Meter   _____

            Lawn Sprinklers (Ground) _____

            Fire incl. hydrants   _____

            Post indicators   _____

            Gas   _____

            Oxygen bulk storage
                  (piping)   _____

            Excavation & backfill   _____

            Paving   _____

            Sodding   _____

CLARE, RANDALL-SMITH & ASSOCIATES LIMITED

-22-

CONSTRUCTION ESTIMATING AND COSTING

8. SERVICES (cont'd)

   (b) PLUMBING & DRAINS (cont'd)

      (10) COMMENTS _____

                          _____

                          _____

                          _____

APPENDIX C

8. SERVICES (cont'd)

   (c) HEATING, VENTILATION AND AIR CONDITIONING

      (1) GENERAL

         Heating System _____

                                      _____

                                      _____

                                      _____

         Cooling System _____

                                      _____

                                      _____

                                      _____

         Ductwork _____

                                      _____

                                      _____

                                      _____

CLARE, RANDALL-SMITH & ASSOCIATES LIMITED

-24-

CONSTRUCTION ESTIMATING AND COSTING

8. SERVICES (cont'd)

   (c) HEATING, VENTILATION AND AIR CONDITIONING (cont'd)

      (2) BOILER ROOM

New boiler type	_____
firing	_____
horse power	_____
hot water temp.	_____
low pressure steam	_____
high pressure steam	_____
Existing boiler	_____
Tunnel system	_____
Ric-wil system	_____
Oil tanks	_____
Breeching	_____
Stacks	_____
Condensate pump & tank	_____
Boiler feed pumps	_____
Oil pumping & heating set	_____
Oil tracing	_____
Blow down tank	_____
Bailey Meter for boiler	_____
De-aerator	_____
Expansion tanks	_____
Convertors	_____
Hot water pumps	_____
Boiler chemical treatment	_____
Water softener for boilers	_____
Master Control centre	_____
Other	_____
	_____
	_____

CLARE, RANDALL-SMITH & ASSOCIATES LIMITED
-25-

APPENDIX C

8. SERVICES (cont'd)

   (c) HEATING, VENTILATION AND AIR CONDITIONING (cont'd)

     (3) PIPING SYSTEMS

Steam H.P.	_____
Steam M.P.	_____
Steam L.P.	_____
Condensate H.P.	_____
Condensate M.P.	_____
Condensate L.P.	_____
Hot water-high temperature	_____
Hot water-low temperature	_____
Oil	_____
Gas (natural)	_____
Condenser water	_____
Chilled water	_____
Ric-wil system	_____
Chemical system	_____
Water softener system	_____
Induction system	_____
Glycol system	_____
Snow melting system	_____
Refrigeration (Direct expansion)	_____
Other	_____

CONSTRUCTION ESTIMATING AND COSTING

8. SERVICES (cont'd)

   (c) HEATING, VENTILATION AND AIR CONDITIONING (cont'd)

      (4) SPECIAL SYSTEMS

          Kitchen hook ups

          Sterilizer hook ups including brass vents

          Owner equipment hook ups

          Tunnel racks and supports

          Other

CLARE, RANDALL-SMITH & ASSOCIATES LIMITED

-27-

APPENDIX C

8. SERVICES (cont'd)

   (c) HEATING, VENTILATION AND AIR CONDITIONING (cont'd)

      (5) EQUIPMENT - Heating & Cooling

- Expansion joints or loops   _____
- Chilled water pumps   _____
- Condenser water pumps   _____
- Cooling tower   _____
- Air cooled condenser   _____
- Refrigeration Machines   _____
- Fan Coil units   _____
- Window induction units   _____
- Wall fin radiation   _____
- Convectors   _____
- Force-flow heaters   _____
- Unit heaters   _____
- Heating coils   _____
- Armstrong humidifiers   _____
- Electric heating coils   _____
- Pressure reducing stations   _____
- Equalizing pressure valves   _____
- Steam metering equipment   _____
- Condensate metering equipment   _____
- Door Heaters   _____
- Fresh Air Make-Up-Direct Gas Fired _____
- Gas Fired Heaters   _____
- Infra Red Heaters   _____
- Other   _____

CLARE, RANDALL-SMITH & ASSOCIATES LIMITED
-28-

CONSTRUCTION ESTIMATING AND COSTING

8. SERVICES (cont'd)

    (c) HEATING, VENTILATION AND AIR CONDITIONING (cont'd)

        (6) EQUIPMENT Air Handling

            Sanitary exhaust _____

            Supply fans _____

            Return fans _____

            Miscellaneous exhaust fans _____

            Cooling coils _____

            Spray coil humidifiers _____

            Pan humidifiers _____

            Grilles _____

            Diffusers _____

            Mixing boxes _____

            Filters _____

            Explosion Proof Units _____

            Terminal Re-heat boxes _____

            Other _____

_____

_____

_____

_____

_____

_____

_____

_____

_____

_____

APPENDIX C

8. SERVICES (cont'd)

   (c) HEATING, VENTILATION AND AIR CONDITIONING (cont'd)

      (7) SHEET METAL

          Sanitary exhaust duct   _____

          Low pressure supply duct   _____

          Low pressure return duct   _____

          High pressure supply duct   _____

          High pressure return duct   _____

          Miscellaneous exhaust ducts _____

          Stainless Steel   _____

          Plastic   _____

          Fibreglass   _____

          Black iron   _____

          Fume hoods   _____

          Pre-manufactured plenums   _____

          Outside louvres   _____

          Fire dampers   _____

          Balancing dampers   _____

          Automatic dampers   _____

          Special Systems   _____

          Garage exhaust   _____

                                                 _____

                                                 _____

                                                 _____

CLARE, RANDALL-SMITH & ASSOCIATES LIMITED

CONSTRUCTION ESTIMATING AND COSTING

8. SERVICES (cont'd)

   (c) HEATING, VENTILATION AND AIR CONDITIONING (cont'd)

      (8) SUB CONTRACTS

           Insulation

           Excavation & backfill

           Concrete bases

           Catwalk and gratings

           Starters and wiring

           Painting

           Hoisting

           Isolation

           Automatic controls

           Other

CLARE, RANDALL-SMITH & ASSOCIATES LIMITED

APPENDIX C

8. SERVICES (cont'd)

   (c) HEATING, VENTILATION AND AIR CONDITIONING (cont'd)

     (9) MISCELLANEOUS

        Permits   _____

        Dept. of Labour Inspection   _____

        Testing and clean up   _____

        Boil out systems   _____

        Sheet metal balance   _____

        Incidentals   _____

        Sleeves   _____

        Tag & directory   _____

        Cutting and patching   _____

        Misc. steel   _____

        Access doors   _____

        Cash allowances   _____

        Freight or cartage   _____

        Temporary heating   _____

        Shack heat light & phone   _____

        Contingencies   _____

                                                       _____

                                                       _____

                                                       _____

                                                       _____

                                                       _____

                                                       _____

CLARE, RANDALL-SMITH & ASSOCIATES LIMITED

CONSTRUCTION ESTIMATING AND COSTING

8. SERVICES (cont'd)

   (c) HEATING, VENTILATION AND AIR CONDITIONING (cont'd)

     (10) COMMENTS _____

          _____
          _____
          _____

CLARE, RANDALL-SMITH & ASSOCIATES LIMITED
-33-

APPENDIX D

The purpose of these two forms is to enable costs to be recorded and controlled during the design development and contract document stages. As an element is designed and costed, or if subsequent revisions are made to it, the cost, together with an appropriate description, is entered on the Cost Plan Check form. A separate form is used for each element, and in some instances, since several alternative designs might be considered for an element, there can be a number of forms for a single element. Each form is numbered consecutively, but only those which are accepted by the design team and signed by the designer are recorded on the Cost Summary. The element target cost given in the cost plan is shown as a reference, together with the most recently updated cost for the element incorporating revisions made since the cost plan was prepared.

As a cost check is approved the amount shown on the Cost Plan Check form is entered onto the Cost Summary. The summary shows, in chronological order, the numbers of the Cost Plan Checks, the elements to which they apply, the amount of each cost check, the revision, if any, to the design contingency, the revised total budget upon approval of the cost check, and the date when it was approved.

APPENDIX D

 Helyar & Associates
Helyar, Rae, Mauchan & Hall Limited
Chartered Quantity Surveyors · Construction Consultants

Project No. _____

COST PLAN CHECK

Project _____ No. _____

Element _____ Date _____

Description: _____
_____
_____
_____
_____
_____
_____
_____
_____
_____
_____
_____
_____
_____
_____
_____
_____
_____
_____

Element Target Cost $ _____
Previous Revised Element Cost $ _____
New Revised Element Cost $ _____

Decision: _____
_____
_____
_____
_____

Signed _____
Architect/Engineer

CONSTRUCTION ESTIMATING AND COSTING

Helyar & Associates
Helyar, Rae, Mauchan & Hall Limited
Chartered Quantity Surveyors · Construction Consultants

Project No. _____

COST SUMMARY

Project _____   Page _____

Check No.	Element	Change	Revised Contingency	Revised Total	Date

Bibliography

Bathurst, Peter E., and David A. Butler: *Building Cost Control Techniques and Economics,* William Heinemann Ltd., London, 1964.

Canadian Institute of Steel Construction and Canadian Steel Industries Construction Council: *A Project Analysis Approach to Buildings,* Toronto, 1976.

Dell'Isola, Alphonse J.: *Value Engineering in the Construction Industry,* Construction Publishing Company, Inc.

Dent, Colin: *Construction Cost Appraisal,* George Goodwin Limited, London, 1975.

Department of Health, Education and Welfare: *Life Cycle Budgeting and Costing,* Washington, Vol. I, 1975; Vol. II, 1967; Vol. III, 1976; Vol. IV, 1976.

Ferry, Douglas J.: *Cost Planning of Buildings,* London, 1964.

Hutton, G. H. and A. D. G. Devonald, eds.: *Value in Building,* Applied Science Publishers Ltd., London, 1974.

Kent, Frederick C. and Maude E.: *Compound Interest and Annuity Tables,* McGraw-Hill Book Company, New York, 1963.

Lichfield, Nathaniel: *Economics of Planned Development,* The Estates Gazette Ltd., London, 1974.

Ministry of Public Building and Works: *Cost Control in Building Design,* Her Majesty's Stationery Office, London, 1968.

Nisbet, James: *Estimating and Cost Control,* B. T. Batsford Ltd., 1968.

Steacey, Richard: *Canadian Real Estate: How to Make It Pay?,* Maclean-Hunter, Toronto, 1970.

Stone, P. A.: *Building Design Evaluation: Costs-in-Use,* E. & F. N. Spon, London, 1967.

Strung, Joseph: *Introduction to Modern Real Estate Investment Analysis and Valuation,* Strung Real Estate Limited, Toronto, 1976.

Index

Access to building, 150; to site, 15
Accommodation, alternative, 147; availability of, 66; cost per unit, 29; owner's requirement, 50; temporary, 147
Accountants, 65
Accounting fees, 68; records, 144
Accumulative rate, 89
Accuracy of estimates, 1, 28-32, 50, 52; of quotations, 27; responsibility for, 53
Adjoining buildings, 15
Administrative costs for development, 69-70
Advertising cost, 70; value of building, 126, 150
Agreement, between client and architect, 2
Agreements, lease, 73, 78; mortgage, 76, 78
Air conditioning, effect of building shape on, 16
All risks insurance, 69
Alteration cost, 31, 74, 146-150; work, 66, 146-150
ALTERATIONS, 35, 55, 170, 190
Alternative, accommodation, 147; designs, 28, 58
Alternatives, comparison of, 137; ranking of, 79, 101, 133, 136-138, 144, 151-153
Amenities, lack of in an existing building, 150; loss of, 147
Amortization, 76, 91; tables, 79, 93
Amount of one dollar, 82; per annum, 84
Analysis, cost, 29, 33-48, 50, 51, 53-55, 59, 161; of designs, 32; error, 151-153; life-cycle cost, 125-156; marginal, 103; sensitivity, 152; of tenders, 6; yield, 94-116
Annual, compounding, 76, 83, 91-92; costs, 72, 73, 133, 136, 137, 140-145, 152; equivalent method, 127, 133-136; equivalent worth, 90, 91, 133, 134, 136, 137, 141, 147, 148; income, 88, 90, 91, 100, 102, 110, 113, 120, 121; interest rate, 83, 91, 92; sinking fund, 86, 88-90, 136
Annuity, 85
Anti-pollution devices, 150

Apartment buildings, 29, 66, 72
Appraisal, xii, 67, 68
Appraisers, 4, 27, 65
Approximate quantities method, 31
Architects, vii, viii, x, 1, 2, 5, 19, 22, 27, 34, 48, 49, 52-55, 57, 60, 65, 126; fees, 67
Architectural merit, 3, 141, 150
Art work, 68
Assets, 86, 87, 95, 98, 119, 136
Assumptions, in early design stage, 50; in calculating life-cycle costs, 138, 151, 153; about selling price of a property, 95; in calculating yield, 94-96, 116; reinvestment, 103, 104, 109, 110
Availability of accommodation, 66; of labour and materials, 9, 14-15

Balance of costs, 48, 49; of mortgage, 105, 112
Balconies, 18, 165
BALCONIES AND PROJECTIONS, 35, 165, 181
BASEMENT, 35, 53, 162, 175; cube, 44
Bills of quantities, x-xii
Block plans, 44, 49
Boat docks, 15
Bond, mortgage, 75; performance, 26, 69; purchase agreement, 68
Bonds, 68, 86, 89, 94-96
Borrowing, 75, 79, 137; methods, ranking of, 79
Breakdown of costs, 24, 27, 28, 33, 34, 36, 51, 56, 59, 60; of man hours, 24, 144
Break-even point, 148
Brokerage commission, 70
Broker's mortgage, 75
Budget, 2, 5, 22, 27, 29, 49, 53, 58, 59, 60; estimate, 34, 40; total, 50, 55
Budgeting program, 57
Building, 86; access, 150; cleaning, 73, 143, 144; components, 20, 34, 49, 140-143, 151, 161; cost, 126; design, 5, 13, 56, 57, 60, 65, 142, 145, 146, 156; elements, 19, 34, 36, 40, 48, 50, 51, 55, 60, 161; equipment, 7, 51, 54, 143-146, 150; height, 14, 15, 18, 19, 30, 51, 53, 55; maintenance, 69, 142-144, 146;

229

management, 73, 145; manager, 144; materials, viii, 9, 14, 15, 18, 19, 20, 23, 25-27, 30, 31, 34, 44, 53, 125, 128, 139, 140-146, 150-153, 156, 161; occupancy, 14, 21, 71, 130, 143; occupants, 143, 146, 147, 150; operation, 71, 146; owner, viii, xi, 1-6, 13-15, 18-22, 28, 29, 48-53, 55, 56, 58, 60, 65, 66, 105, 125, 126, 137, 141-144, 153; quality, 14, 15, 29-31, 40, 44, 49-53, 55, 58, 71; repairs, 7; shape, 14-16, 18, 19, 29-31, 36, 40 44, 51, 52, 55, 125; site, 9, 15, 55, 77, 132; size, 14-16, 18, 19, 29-31, 36, 40, 44, 49, 51, 52, 55, 125; systems, 14, 20, 31, 34, 53, 54, 125, 143, 145, 151; type, 13-15, 49, 54, 141, 143, 145

Buildings, apartment, 29, 66, 72; commercial, 14, 76, 77; existing, 66, 67, 71, 144, 147, 149, 150; industrial, 14, 66, 75, 78; institutional, 14, 65; life of, 125, 126, 128, 137, 140-142, 145, 146, 152, 153, 156; multi-storey, 18; office, 2, 4, 66, 72, 110, 135; public, 143, 151, 153; residential, 14, 51, 66

By-laws, planning, 15, 18

Calculators, electronic, 81, 93, 113, 132, 140
Canadian, amortization schedules, 93; corporations, 118; -controlled private corporations, 118, 119; Institute of Quantity Surveyors, xii, 23, 30, 34-36, 157, 161; Metric Practice Guide, 158; mortgages, 92; Standards Association, 158
Canopies, 165
Capital, 99, 104, 126, 150
Capital, cost, 19, 67-71, 96, 97, 99, 100, 103, 106, 109, 112, 119-121, 125-129, 132, 137, 141-153, 156; cost allowance, 119-121, 123; formation, 7; gain, 95, 98, 119; goods, 9; investment, 88; levies, 69; loss, 95; security of, 95; stock, 69; taxes, 69
Capitalization, 67, 71, 97; methods, 96-101, 106, 112; rates, 96, 97, 99, 100, 107
Cash, allowances, 5, 26; flow, 95, 102, 104, 110, 114, 120, 121, 122, 139, 140; flow discounted, 101, 102, 105,
114-115, 121, 140
CEILING FINISHES, 35, 167, 186
Central Mortgage and Housing Corporation, 77
Certificate of payment, xi
Change order form, 59
Changes, in design, 4; in technology, 1, 141; in variables for life-cycle costs, 152, 153; in the work, xi, 3, 5, 20, 27
Circulation space, 19
Cladding, cost, 16, 19; exterior, 16, 19, 32, 44, 53, 58, 147; ratio, 16, 18, 19
Client and architect agreement, 2
Collateral, 95
Column spacing, 53, 147
Columns, 16, 19, 20, 72
Commissions, leasing, 70; real estate, 67
Communication equipment, 73
Community, effect of a building on, 150; interests, 150
Comparisons, of alternatives, 137; of costs, 29, 32, 36, 48, 125-127, 138, 143, 151, 157
Completion of construction, 21, 70, 71, 116, 130, 131
Components, building, 20, 34, 49, 140-143, 151, 161; equipment, 145, 146; life of, 140, 142, 152; repairs, 142, 146, 156; replacement, 142, 143, 146, 156; standard, 20
Compound interest, 80-91
Compounding, 76, 109
Compounding periods, 76, 83, 91-93
Computers, xi, 81, 113, 136, 140, 153
Construction activity, 8; Arctic, 15; completion, 21, 70, 71, 116, 130, 131; contract, 20-22; costs, 1-5, 7, 9, 13-16, 18-22, 27, 29, 30, 33, 34, 48-52, 54, 55, 56, 67-69, 71, 98, 130, 137, 139, 143, 157; details, 20, 30, 146; duration of, 28; economics, vii; industry, vii, ix, 7-13; management, 3, 13, 21, 22, 60, 143; manager, 3, 21, 60; non-residential, 10, 12, 51; postponement of, 67; program, 49; project, 6; residential, 11, 12, 14, 51; schedule, 22; standards, 137; techniques, 1, 13; time, 28, 131, 132, 143; type of, 29-31, 40, 44, 55, 145, 161

230

Consultants, xii, 1-6, 13, 21, 22, 27-29, 34, 49, 50, 53, 55, 56, 60, 65, 68, 69, 71, 77, 78; fees, 67
CONTINGENCIES, 35, 55, 172, 191
Contingency, 28, 51, 52, 55, 58, 59, 71
Contract, cost plus, viii, 21, 22; document stage, 59-60; documents, xi, 6, 20, 52, 57; maintenance, 143; management – see construction management; service, 73, 143; stipulated sum, 5, 13, 20, 22, 28
Contractors, vii-x, 1, 5-7, 9, 13, 14, 20-28, 30, 49, 51, 56, 59, 69, 70; equipment, 18, 23, 25-27, 31; estimate, 6, 25, 27, 28, 56; objectives, 28; overhead, 5, 21; profit, 5, 21, 23, 25, 26; warranty, 69
Co-ordination of tenders, 5
Corporation tax, 117-118
Corporations, 13, 69, 117-122
Corridors, 72
Cost, advice, x-xii; analysis, 29, 33, 34, 48; breakdown, 24, 27, 28, 33, 34, 36, 51, 56, 59, 60; checking, 57-59; commitment by the owner, 21, 22, 28, 50; comparisons, 29, 32, 36, 48, 125-127, 138, 143, 151, 157; control, xi, xii, 1-6, 28-31, 36, 49, 57-62, 125; control, need for, 1-6; control, objectives of, 49, 125; decisions, 55, 58; distribution, 48, 57, 59; elemental analysis, 33-48, 50, 51, 53-55, 59, 161; escalation, 28, 67, 72, 116, 137, 139, 140, 144, 153; forecasting, 5, 48, 60; implications, 20, 60, 126; index, 51; information, 20, 26-29, 34, 48, 50, 142; of labour, viii, 9, 14, 20, 23, 24, 26, 27, 31, 33; of materials, 9, 26, 27, 31, 143, 144, 151; plan, 55-60, 62; plan check form, 225, 226; planning, xi, xii, 48-57, 60-62; per square metre, 13, 19, 29, 30, 33, 40, 44, 48, 50, 54, 55, 161; plus, viii, 21, 22; significance, 34, 48, 152; of site, 15; summary form, 225, 227; target, 57-59; and value, 3, 4, 65, 96, 97
Costing, life, 125; life-cycle, 125-156; life span, 125
Costs, administrative, 69-70; advertising, 70; alteration, 31, 74, 146-150; annual, 72, 73, 133, 136, 137, 140-145, 152; building, 126;

capital, 19, 67-71, 96, 97, 99, 100, 103, 106, 109, 112, 119-121, 125-129, 132, 137, 141-153, 156; construction, 1-5, 7, 9, 13-16, 18-22, 27, 29, 30, 33, 34, 48-52, 54, 55, 56, 67-69, 71, 98, 130, 137, 139, 143, 157; definition of, 30; development, 51, 67-71, 98; distribution of, 48, 57-59; errors in, 151, 152; factors influencing, 7-22; financing, 73, 145, 153; future, 74, 96, 125-127, 133, 137, 139, 141, 142, 144, 151, 152; hard, 67; initial maintenance, 69; initial occupancy, 71; in-use, 125; investors, 95, 96; land, 7, 67; maintenance, 69, 125, 129, 130, 139, 141, 143, 144, 146, 150, 153; moving, 147; negative costs, 150; occupancy, 71, 72, 144, 145; operating, 19, 71-74, 116, 125, 126, 139, 141, 143, 144, 146-150, 152, 153, 156; ownership, 72, 73, 133, 135; periodic, 72, 73, 127, 131, 133, 134, 137, 140, 143, 145-150; of production, 144; of renting, 133, 135; of repairs, 128, 133; replacement, xii, 127, 133, 139, 145, 146; structural, 147; target, 57-59; total, 23, 65, 79, 98, 125, 145; transfer, 96; ultimate, 125; unit, 18, 44
Crew cost method, 24
Cube, of the basement, 44; of the building, 1, 30
Cubic metre method, 30
Cumulative deduction account, 118, 119, 121
Current liabilities, 13

Debentures, 70
Decisions, 29, 32, 50, 55-60, 67, 95, 125, 142, 145, 146, 148, 150
Deeds, trust, 68
Deferred value, 84
Delay, 2, 3, 14, 21, 22, 28, 145
Demand for loans, 77
Demising partitions, 72
DEMOLITION, 35, 55, 171, 190
Demolition, 66, 67, 98, 108, 114, 125, 130, 141, 149, 150
Depreciable, assets, 119; real estate, 120
Depreciated capital cost, 123

Depreciation, 119, 120
Design, alternatives, 28, 58; analysis of, 32; assumptions, 50; of building, 5, 13, 56, 57, 60, 65, 142, 145, 146, 156; changes, 4; consultants, 1-6, 13, 21, 22, 27-29, 34, 50, 55, 56, 71; contingencies, 55, 58, 59; decisions, 32, 55, 56, 58-60; detailed, 55; development stage, 32, 57-59; drawings, 52, 58; economical, 126; of elements, 58, 59; fees, 5, 51, 68, 143; information, 28, 31; mechanical, 18; process, 49, 56, 58, 59; solutions, 28, 58; stage, 34, 49, 125, 145, 146; structural, 18; team, 20, 22, 52, 55-57, 60, 62, 70, 142
DESIGN CONTINGENCY, 35, 55, 172, 191
Designer, 13, 20, 30, 58, 60
Designing by elements, 57, 60
Detailed design, 55; drawings, 27
Detailing, 20, 30, 146
Developer, vii, 2, 4, 65-70, 72, 75, 77-79, 86, 96-105, 107-110, 112, 114, 120, 122, 126, 137
Developer's fee, 68, 69; profit, 104, 123; staff, 69, 73
Development, company, 78, 117, 118, 120-122; costs, 51, 67-71, 98; participation in, 77; phased, 67, 69, 71; real estate, 65-69, 71, 75, 77, 96-107, 109, 110, 112, 114, 117, 120, 122, 137, 151; sale of, 94, 95, 100-103, 105, 107, 109, 110, 112, 119-123; value of, 4, 66, 71, 96
Diminishing balance method, 119
Discount, rate, 90, 127, 128, 135-137, 141, 142, 148, 152, 153; rate, modified, 138-140, 148; tables, 81, 89, 93, 140
Discounted cash flow, 101, 102, 105, 114-115, 121, 140
Discounting, 76, 84, 96, 109, 114, 127, 128, 130, 137, 140, 147; formulae, 80, 82, 84-86, 88-92, 96, 97
Distribution of costs, 48, 57-59
Dividends, 118, 119, 121
Dominion Interest Act, 76, 77, 92
Doors, 53, 54
Drawings, vii, ix, 20-22, 27, 40, 44, 49, 50, 52, 58, 59, 77, 161
Dual rate tables, 89
Duty on imported materials, 25

Early occupancy, 143
Economic, decisions, 150; factors, 141; gain, 146; growth, 67; life of buildings, 141; viability, 65; worth, 66
Effective, interest rate, 92; tax rate, 118
Efficiency, planning, 19, 30
ELECTRICAL, 35, 168, 193-205
Electrical, costs, 36; engineers, 52, 54-55; systems, 18, 147
Electronic calculators, 81, 93, 113, 132, 140
Elemental, breakdown, 34, 51, 56, 59; cost analysis, 33-48, 50, 51, 53-55, 59, 161; method, 32, 33, 49
Elements, 19, 34, 36, 40, 48, 50, 51, 55-60, 161; definition of, 161-172; design of, 58, 59; designing by, 57, 60; function of, 34, 161; list of, 34-35, 162-172; measurement of, 23, 25, 28, 34, 44, 53-56
Elevations, 161
Elevators, 18, 73
ELEVATORS AND ESCALATORS, 35, 54, 166, 184
Engineer, electrical, 52, 54-55; mechanical, 52, 54-55; structural, 20, 52-54
EQUIPMENT, 35, 168, 188
Equipment, building, 7, 51, 54, 143-146, 150; communication, 73; components, 145, 146; contractors', 18, 23, 25-27, 31; failure of, 143, 144, 146; hoisting, 18; life of, 145-146; material handling, 147; rental, 26; stand-by, 143
Equity, 99, 100, 106, 112, 137
Equity financing, 13
Error analysis, 151-153
Errors, compensating, 31; in costs, 151, 152; mathematical, 44; in quantities, x, xi; in unit prices, 31
Escalation, 28, 67, 72, 116, 137, 139, 140, 144, 153
ESCALATION CONTINGENCY, 35, 55, 172, 191
Estimate, accuracy of, 1, 28-32, 50, 52; contractor's, 6, 25, 27, 28, 56; design consultant's, 1, 2, 15, 27-29, 31-33, 36, 40, 44, 50-56, 60, 62, 157, 161; final, 2, 62; first, 50, 52;

232

preliminary, 15, 28, 36, 44, 60;
pre-tender, 36
Estimating, construction costs, 29,
31, 33, 44, 48, 49, 53, 54, 56, 60, 144;
data, 27; department, 5, 24, 28; the
life of buildings, 141; methods,
23-33; preliminary, 44; speed of,
28-30, 32
Estimator, xi, 6, 20, 24, 25, 26
Existing, buildings, 66, 67, 71, 144,
147, 149, 150; leases, 70
EXTERIOR CLADDING, 35, 53, 163, 179
Exterior, cladding, 16, 19, 32, 44, 53,
58, 147; walls, 16, 34, 54, 58, 72,
157-159
EXTERIOR DOORS AND SCREENS, 35,
53, 164, 181

Factors, influencing cost, 7-22; of
production, 126
Factory, 18, 72, 126, 144, 147, 148
Failure of equipment, 143, 144, 146
Fast track, 3, 21, 60
Faulty materials, 146
Feasibility studies, 2, 49, 65-74, 78,
96, 138, 143, 150, 151
Federal government, 117, 118
Federal Income Tax Act, 117, 119
Fees, accounting, 68; appraisal, 68;
architect's, 67; audit, 73;
construction management, 21;
consultants, 67; design, 5, 51, 68,
143; developer's, 68, 69; finders, 70;
interim lender commitment, 70;
legal, 68, 73; long term lender
standby, 70; professional, 5; project
management, 68; registration, 70;
special consultants', 68;
underwriters', 68
Final account, xi; estimate, 2, 62
Finance costs, 73, 145, 153; interim,
70, 93
Financial, decision, 2, 50; factor, 141;
limitations, 126; sources, 137;
statements, 78; supervision, xii
Financing, 13, 28, 75-79, 93, 99, 105,
106, 122, 137, 145, 151; costs, 153;
permanent, 70
Finder's fee, 70
Finishes, 14, 22, 44, 48, 54, 70, 145,
146; floor, 127, 129, 134, 146; roof,
18, 127, 146; soffit, 18; value of, 54;
wall, 54
Fire insurance, 145

First, estimate, 50, 52; mortgage, 77
Fittings, 51, 54, 147
FITTINGS AND EQUIPMENT, 35, 54,
168, 187
FITTINGS AND FIXTURES, 35, 168, 187
Fixed capital formation, 7, 8
Fixtures, 51, 54; lighting, 169, 196
Floor area, gross, 1, 18, 29, 33, 40, 44,
48, 50, 54, 72, 157, 159, 161;
measurement of, 157-160; net, 18,
49, 50, 72
Floor bearing capacity, 147
FLOOR FINISHES, 35, 167, 185
Floor finishes, 127, 129, 134, 146;
slabs, 53
Floors, 44
Footprint of the building, 53, 162
Forecasting costs, 5, 48, 60
Foreclosure of mortgage, 77
Formulae used for discounting, 80, 82,
84-86, 88-92, 96, 97; for calculating
the internal rate of return, 106
Formwork, 20
Foundations, 15, 18, 44
Fuel, 73, 139, 144
Full-floor occupancy, 72
Function, of the building, 50, 126, 145;
of elements, 34, 161
Furnishings, 30, 68
Furniture, 68, 147
Future, alterations, 74, 147, 148; costs
(expenses), 74, 96, 125-127, 133,
137, 139, 141, 142, 144, 151, 152;
income, 96, 97, 106; value, 81, 82,
84, 91, 109, 130, 132, 133, 136;
worth of one dollar, 82-86, 132

Garage, parking, 14, 29
Garbage, removal, 73
Gardening, 73
GENERAL SITE DEVELOPMENT, 35,
170, 189
GLAZED PARTITIONS AND DOORS, 35,
166, 183
Government, departments, 2, 153;
federal, 117, 118; offices, 151;
policy, 147; provincial, 69
Gross, fixed capital formation, 7, 8;
floor area, 1, 18, 29, 33, 40, 44, 48,
50, 54, 72, 157-159, 161; floor area,
measurement of, 157-160; national
product, 7, 8
Ground floor, 18, 19, 72; plan, vii, viii

233

HEAD OFFICE OVERHEAD AND PROFIT, 35, 171
Heat, gain, 19; losses, 19
Heating systems, 16, 128, 134, 139, 144, 152
HEATING, VENTILATION AND AIR CONDITIONING, 35, 169, 215-224
Height of building, 14, 15, 18, 19, 30, 51, 53, 55
High-rise structures, 18, 143
Hoisting equipment, 18
Holding deposit, 78
Hospitals, 9, 13, 29, 72, 144, 147, 151
Hourly labour rate, 25

Imported materials, 25
Income, 4, 6, 66, 69, 71-73, 75, 77-79, 87-91, 94-97, 99, 100, 102, 105, 106, 108-110, 112-114, 116, 118-121, 123, 127, 139, 145, 147; annual, 88, 90, 91, 100, 102, 110, 113, 120, 121; during construction, 71; future, 96, 97, 106; investment, 118; loss of, 147; negative, 104-105, 106; negative taxable, 123; perpetual, 87, 88, 90, 136; potential, 126, 145, 151; -producing building, 66, 71-72, 88, 119, 137, 147; received for a limited time, 88, 90; rental, 70, 118, 119; security of, 95; Tax Act, 117, 119; taxable, 118, 119, 121, 123
Industrial buildings, 14, 66, 75, 78
Industries, manufacturing, 9, 13
Inflation, 1, 4, 7, 65, 80, 95, 116, 127, 136-140
Inflation rate, 138, 139
Information, cost, 20, 26-29, 34, 48, 50, 142; design, 28, 31
Initial, maintenance costs, 69; occupancy costs, 71
Input-cost index, 51
Instability, in the construction industry, 7-8; of the economy, 9
Institutional, buildings, 14, 65; owner, 65
Insulation, 18
Insurance, 26, 69, 73, 145; all risks, 69; fire, 145; company, 89; mortgage, 69; premiums, 69, 85; title, 67
Interest, 76, 79-83, 86, 87, 91, 93, 99; and annuity tables – see Discount tables; compound, 80-91; rate, 7, 50, 70, 76-86, 88-90, 92, 94, 96, 137,

138, 145; simple, 80, 81, 87, 88
Interim, finance, 70, 93; lender commitment fees, 70
Interior, doors, 53, 54; finishes, 48, 54, 146; partitions, 19, 53, 54; planning, 16, 48, 49, 52, 125
INTERIOR FINISHES, 35, 54, 167, 185
INTERIOR PARTITIONS AND DOORS, 35, 53, 54, 165, 182
Internal communication systems, 73; rate of return, 106-110, 112-114, 120-124, 136, 137, 140, 149
Interstitial space, 147, 148
Investing, 66, 75, 80-89, 91, 96, 103, 137
Investment, dealers, 75; decisions, 95; income, 118; length of, 80-86, 88, 99, 103; management, 96; return on, 66, 68, 79, 87, 92, 94, 97
Investments, 4, 9, 66, 71, 75, 80, 81, 85-91, 94-96, 98-100, 103, 106, 109, 116, 137
Investors, xii, 66, 81, 83, 87, 88, 94-96, 117

Labour, constants, 24, 25, 27; cost, viii, 9, 14, 20, 23, 24, 26, 27, 31, 33; cost index, 51; productivity, 14, 25; quality, 14, 25; quantity, 23; rates, 14, 25; requirements, 25; shortages, 14
Land, 67, 69, 86, 87, 98, 108, 136, 143; cost, 7, 67; current value, 67; purchase, 68, 130, 131; rental, 73; surveys, 67; transfer charges, 68; transfer taxes, 67; value of, 67, 98
Landscaping, 150
Lawyers, 65
Lease, agreements, 73, 78; renewals, 67; takeovers, tenant, 70; term of, 67
Leases, 68, 70, 72
Leasing, 70, 73; agent, 70; arrangements, 72; commissions, 70; practices, 67
Legal fees, 68, 73
Lender – see Mortgagee
Length of investment, 80-86, 88, 99, 103
Leverage, 75, 79, 106
Levies, 69
Life, of buildings, 125, 126, 128, 137, 140-142, 145, 146, 152, 153, 156; of

components, 140, 142, 152; costing, 125; -cycle cost assumptions, 138, 151, 153; cycle costing, 125-156; of equipment, 145-146; of materials, 128, 139, 140, 142, 152, 153, 156; span costing, 125
Lighting fixtures, 169, 196
Lighting and power distribution, 169
Limited time, income received for, 88, 90
List of elements, 34-35, 162-172
Loan, commitment, 78; terms of, 76
Loans, 13, 69, 70, 75-79, 86, 93, 95, 99; N.H.A., 76-77
Location of the building, 14, 15, 49, 51, 141
Long, -life buildings, 141, 142; -term financing, 77; -term lender standby fees, 70; -term loans, 76
Loss, of amenities, 147; of income, 147; of production, 147; tax, 123; terminal, 119
LOWEST FLOOR CONSTRUCTION, 35, 44, 162, 176
Low-rise structure, 143
Low-risk bonds, 89
Lump sum, 21, 26, 36, 44, 54, 55

Maintenance, building, 142-144; contract, 143; costs, 69, 125, 129, 130, 139, 141, 143, 144, 146, 150, 153; operations, 143, 144; planned, 143; policy, 142, 143; pool, 73; preventative, 73; remedial, 144; staff, 143; tasks, 143; work, 69, 144
Management, building, 73, 145; construction, 3, 13, 21, 22, 60, 143; of investments, 96; office, 118; property, 73; staff, 118
Manager, building, 144; construction, 3, 21, 60; project, vii, 60, 69; property, 142
Manufacturers, 4, 141, 142
Manufacturing industries, 9, 13
Market, analysts, 65; conditions, 48, 51, 71; stock, 95; studies (surveys), 66-68, 71; value, 3
Masonry, 34
Material, cost index, 51; handling equipment, 147
Materials, viii, 14, 15, 18, 19, 20, 23, 25, 27, 30, 34, 44, 53, 125, 141, 142, 145, 150, 151, 161; cost of, 9, 26, 27, 31, 143, 144, 151; delivery of, 25; faulty, 146; imported, 25; life of, 128, 139, 140, 142, 152, 153, 156; local, 15; new, 142; quality of, 20, quantities of, 23; repairs of, 127, 142, 143, 156; replacement of, 127, 142, 145, 156
Mathematical errors, 44
Maximum upset price, 21
Measurement, of elements, 23, 25, 28, 34, 44, 53-56; of gross floor area, 157-160; of quantities, 162-172
Measurements, definition of, 23, 30
Measurer, viii
Measuring surveyor, viii-xi
Mechanical costs, 36; design, 18; engineer, 52, 54-55; spaces, 19, 72; systems, 18, 147
Medical, centres, 66; staff, 72, 144
Modified design, 5; discount rate, 138-140, 148; internal rate of return, 109-110
Money, 1, 22, 48, 49, 60, 80, 81, 92, 95, 125, 142, 153; time and, 80-93; value for, 1, 48, 49, 60, 125, 153; value of, 80, 95
Monthly, compounding, 92; interest-rate, 83, 92; payments, 76, 92
Mortgage, agreement, 76, 78; amortization tables, 79, 93; appraisals, xii, 68; balance remaining, 105, 112; bonds, 75; brokers, 75; company, 68-70; discharge, 77; first, 77; foreclosure, 18; insurance, 69; interest rates, 50, 137, 145; lenders, 75; loan, 70, 76, 99; open, 76; penalty, 77; prepayment privileges, 76; principal, 76, 79, 92, 99, 100, 105, 112, 123; repayment, 69, 76, 78, 79, 85, 92, 99, 100, 110, 112, 122, 123; second, 77; term, 76, 77
Mortgagee, 76-78, 94, 99
Mortgages, 67, 75-78, 91, 92, 94, 99, 100, 105, 106, 112, 122
Mortgagor, 76, 77, 94
MOVABLE PARTITIONS AND DOORS, 35, 166, 182
Moving costs, 147
Multiple-rate methods, 29, 31, 33
Multipliers for mortgage payments, 79
Multi-storey buildings, 18

Municipal, levies, 69; politicians, 65; services, 69
Municipalities, 15, 69

Natural constant, 79
Negative, cost, 150; incomes, 104-105, 106; taxable income, 123
Net, annual income – see Income, annual; floor area, 18, 49, 50, 72; to gross floor area ratio, 18, 19, 50; present value, 101-107, 120, 127
New materials, 142
N.H.A. loans, 76-77
Noise abatement, 150
Nominal interest rate, 92, 94
Non-resident corporations, 118
Non-residents, 117
Non-residential construction, 10, 12, 51
Non-union contractors, 14
Non-wasting assets, 87, 98, 136
NORMAL FOUNDATIONS, 34, 162, 175

Objectives, of the building owner, 126, 150; of the contractor, 28; of cost control, 49, 125; of the investor, 95
Obsolescence, 141
Occupancy of the building, delay in, 21; early, 143; full floor, 72; time of, 71, 130; type of, 14
Occupancy costs, 71, 72, 144, 145
Occupants of the building, 143, 146, 147, 150
Office, buildings, 2, 4, 66, 72, 110, 135; management, 118
Offices, government, 151; site, 26
Ontario tax returns, 118
Open mortgage, 76
Operating costs, 19, 71-74, 116, 125, 126, 139, 141, 143, 144, 146-150, 152, 153, 156
Operating staff, 73
Operation of the building, 71, 146
Operations, site, 23, 24
Ordinary annuity, 85
Outline specification, 52, 56, 173-224
Output cost index, 51
Overexpenditure on capital costs, 153
Overhangs, 165
Overhead, contractor's, 5, 21; expense, 5; head office, 5, 26; site, 25, 26

OVERHEAD AND PROFIT, 35, 55, 171, 191
Overtime, 147
Owner – see building owner; cost commitment, 21, 22, 28, 50; debt, 137
Owner's equity, 99, 100, 106, 112; objectives, 126, 150; requirements, 126
Ownership costs, 72, 73

Parking, 69, 150; garage, 14, 29; levies, 69; public, 69; structure, 4
Partial payment factor, 90
Participation clause, 21
Participation in a development, 77
Partitions, 19, 44, 53, 54, 147; concrete, 36; demising, 72; dividing, 36; movable, 147; solid, 147; structural, 36; tenant, 70
Percentage, addition for contingency, 52, 172; breakdown of costs, 36; of finishes, 54, 168; markup, 21; addition for overhead and profit, 55; of receipts as rent, 72
Performance bonds, 26, 69
Periodic costs, 72, 73, 127, 131, 133, 134, 137, 140, 143, 145-150; redecorations, 73, 110; repairs, 73, 110
PERMANENT PARTITIONS AND DOORS, 35, 36, 54, 165, 182
Perpetual income (in perpetuity), 87, 88, 90, 136
Phased development, 67, 69, 71
Physical life, of buildings, 141, 145; of equipment, 145-146
Planned maintenance, 143
Planning, by-laws, 15, 18; efficiency, 19, 30; interior, 16, 48, 49, 52, 125; solutions, 145
Plans, 44, 49, 161
PLUMBING AND DRAINS, 35, 169, 206-214
Pool maintenance, 73
Portmanteau unit-price, 31
Positive cash flow, 139
POST-CONTRACT CONTINGENCY, 35, 55, 172, 191
Postponement of construction, 67
Potential income, 126, 145, 151
Preliminary drawings, 44, 52, 58; estimate, 15, 28, 36, 44, 60

Premiums, insurance, 69, 85
Prepayment privileges, 76
Present value, 84, 86, 96-100, 102, 104-106, 108, 109, 114, 117, 128, 130, 132-134, 136-141, 150; factors, 114; method, 126-133, 136; ratio, 103-107, 113
Present worth, 84; of one dollar, 83, 84, 87-90, 97, 148
Prestige, considerations of, 65, 66, 95, 126, 150
Pre-tender estimate, 36
Preventative maintenance, 73
Price, quotations, 27, 28; unit, viii, xi, 1, 6, 15, 23, 25-28, 30, 31, 44, 48, 53, 56, 161
Principal, 76, 79-82, 91, 92, 99, 100, 105, 112, 123
Private, clients, 2; corporations, 118, 119
Probability, of life expectancy, 142; of reduction in operating costs, 152
Production cost, 144; factors of, 126; loss of, 147
Productivity, 14, 25, 51
Products, custom-made, 20
Profit, contractor's, 5, 21, 23, 25, 26; developer's, 104, 123; margins, 7
Profitability, 5, 78, 150
Pro-forma for an office building, 110
Program, 49, 50
Program stage, vii, 49, 53
Progress payments, x, xi, 28
Project management fee, 68; manager, vii, 60, 69
Projecting balconies, 18
Projections, 18, 19
Property, company, 68, 75; management, 73; manager, 142
Protection of furniture and fittings, 147
Provincial, governments, 69; tax abatement, 120; taxes, 117, 118, 120
Provision for future alterations, 147, 148
Public, buildings, 143, 151, 153; corporations, 118; parking, 69; relations, 70
Pumping costs, 16

Quality, of the building, 14, 15, 29-31, 40, 44, 49-53, 55, 58, 71; of drawings, 22; of labour, 14, 25; of materials, 20
Quantities, ix, xii, 23, 27, 44, 48, 54, 56; guessing, 54; measurement of, 162-172
Quantity, survey method, 23, 31-34; surveyors, vii, x, xi, xii, 27, 65
Quarterly compounding, 91, 92
Quebec tax returns, 118
Quotations, 27, 28

Ranking, of borrowing methods, 79; of developments, 101, 103; of life-cycle costs, 133, 136-138, 144, 151-153
Rates, accumulative, 89; capitalization, 96, 97, 99, 100, 107; discount, 90, 127, 128, 135-137, 141, 142, 148, 152, 153; effective interest, 92; effective tax, 118; inflation, 138, 139; interest, 7, 50, 70, 76-86, 88-90, 94, 96, 137, 138, 145; labour, 14, 25; modified discount, 138-140, 148; mortgage interest, 50, 137, 145; nominal, 92, 94; reinvestment, 103, 104, 109; remunerative, 89; rental, 4, 72, 116, 135, 136; of return, 76, 79; stated, 92; target, 106-108, 112-114, 123; tax, 117, 118, 120, 122; vacancy, 86; yield, 75, 94, 96-109, 113, 114, 116, 121, 122, 124
Ratio, between capital and operating costs, 145, 152; between element and gross floor area, 16, 18, 19, 44, 48, 54, 161; between net and gross floor areas 18, 19, 50; of participation in savings, 21; present value, 103-107, 113
Real estate, commissions, 67; developments, 65-69, 71, 75, 77, 96-107, 109, 110, 112, 114, 117, 120, 122, 137, 151; investments, 95; investors, 94; taxes, 69, 72, 73
Realtors, 65
Rebuilding, 66, 141, 149
Recording costs during the design stage, 225
Redecorating, 73, 110, 145
Registration fees, 70
Reinforced concrete, 20, 31, 53
Reinvestment, 81; assumptions, 103, 104, 109, 110; rate, 103, 104, 109
Remedial action, 57, 59; maintenance,

237

144
Remunerative rate, 89
Renewal of leases, 67
Renovations, 21, 66, 125, 149, 150, 156
Rent controls, 96
Rental, building, – see
  Income-producing building;
  equipment, 26; income, 70, 118,
  119; land, 73; rates, 4, 72, 116, 135,
  136; terms, 78
Renting, 4, 125, 133, 135, 136
Rents, 4, 15, 50, 67, 71-73, 85, 133
Repairs, of buildings, 7; of
  components, 142, 146, 156; costs
  of, 128, 133; of materials, 127, 142,
  143, 156; periodic, 73, 110
Repayment, of loans, 86; of mortgage,
  69, 76, 78, 79, 85, 92, 99, 100, 110,
  112, 122, 123
Replacement, of components, 142,
  143, 146, 156; costs, xii, 127, 133,
  139, 145, 146; cycle, 130; of
  materials, 127, 142, 145, 156
Requirements, labour, 25; owner's,
  126
Residential construction, 11, 12, 14,
  51
Residual value, 99
Responsibility, for accuracy of
  estimate, 53; for cost control, 60
Restaurants, 72
Retail stores, 72
Retained earnings, 69
Return on investment, 66, 68, 79, 87,
  92, 94, 97 (see also Yield)
Revenue, increased, 126
Re-zoning applications, 68
Roads, 9, 15
ROOF CONSTRUCTION, 35, 163, 178
Roof construction, 18
ROOF FINISH, 35, 163, 179
Roof finish, 18, 127, 146
Roof slabs, 53
Roofs, 18, 44, 133, 134, 139
Routine servicing, 143
Rural sites, 15

Safety, standards of, 150
Salaries, 70, 150
Sale of a development, 94, 95,
  100-103, 105, 107, 109, 110, 112,
  119-123, 125, 153
Sales taxes, 25

Salvage, 150
Salvage value, 139
Saving in alteration costs, 148
Savings, 80; shared, 21
Schematics stage, 52-58, 60
Schematics stage, 52-58, 60
School board, 29
School building program, 40
Schools, 9, 29, 147, 151
Second mortgage, 77
Secrecy in tendering, x
Securities, 94
Security, 73; of capital, 95; of income,
  95; personnel, 144
Selected list of tenderers, 6
Selling price of a property, 95
Semi-annual compounding, 76, 91-93
Sensitivity analyses, 152
Service, agencies, 9; contracts, 73,
  143
SERVICES, 35, 36, 54, 168, 189, 193-224
Services, municipal, 69
Servicing, routine, 143
Sewers, 9, 69
Shape of the building, 14-16, 18, 19,
  29-31, 36, 40, 44, 51, 52, 55, 125
Shared savings, 21
Shares, 75
Sheet piling, 15
Shopping centres, 66, 72
Shoring, 15
Shortages, of labour, 14
Short-life buildings, 141, 142
Significance of cost, 48
Significant cost, 34, 152
Simple interest, 80, 81, 87, 88
Simplicity of drawings, 20
Single rate methods, 29-31, 33, 34, 51;
  rate tables, 89; storey building, 18
Sinking fund – see Annual sinking
  fund
Sinking fund policy, 89
SITE DEVELOPMENT, 35, 55, 170, 189
SITE OVERHEAD, 35, 171
SITE SERVICES, 35, 170, 190
Site, access, 15; building, 9, 15, 55, 77,
  132; conditions, 25, 48, 51; cost, 15;
  development, 55; offices, 26;
  operations, 23, 24; overhead, 25,
  26; staff, 24, 25; works, 30, 51
Size of the building, 14-16, 18, 19,
  29-31, 36, 40, 44, 49, 51, 52, 55, 125
Sketch drawing stage, 31

238

Sketch drawings, 40, 44, 52
Slab on grade, 18
Sleeping accommodation, 15
Small business deduction, 118-120
Snow removal, 73
Soffit finish, 18
Soft costs – see Development costs
Soil conditions, 15, 51, 53
Soil tests, 68
Space, useable, 29, 30
Special consultants, 65, 68
Special consultants' fees, 68
SPECIAL FOUNDATIONS, 35, 162, 176
Special structures, 9
Specifications, 5, 20-22, 26, 27, 40, 49, 59, 60, 77; outline, 52, 56, 173-224
Speed of construction, 70; of estimating, 28-30, 32
Square foot method – see Square metre method
Square metre, cost, 13, 19, 29, 30, 33, 40, 44, 48, 50, 54, 55, 161; method, 29, 30, 33, 50, 51
Staff, building, 145; contractor's, xii, 26; developer's, 69, 73; maintenance, 143; management, 118; medical, 72, 144; morale, 150; operating, 73; site, 24, 25; training, 71; turnover, 150
Stage, contract document, 59, 60; design, 34, 49, 125, 145, 146; design development, 32, 57-59; program, vii, 49, 53; schematics, 52-58, 60; sketch drawing, 40, 44, 52
STAIRS, 35, 54, 166, 183
Stairs, 18, 147
Standard components, 20
Standardization, 13
Standards of comfort and safety, 150
Stand-by equipment, 143
Stipulated sum, x, 5, 13, 20-22, 28
Stock, capital, 69
Stock market, 95
Storage, 15, 19
Storey heights, 18, 30, 147
Structural, alterations, 147; costs, 147; design, 18; engineer, 20, 52-54; frames, 36; partitions, 36; requirements, 147; steel, 53; systems, 32
STRUCTURE, 35, 40, 53, 162, 176
Structures, high-rise, 18, 143; low-rise, 143; parking, 4; special, 9

Student residence, 40
Sub-contractors, xi, 5, 6, 13, 14, 21, 22, 26-28
Sub-elements, 40, 53, 54
SUBSTRUCTURE, 34, 36, 53, 162, 175
Sub-trade prices, 28
Suburban sites, 15
Superintendent, 24
Supervision, 26; financial, xii
Suppliers, 25, 27, 28
Supplies, 14, 73, 144
Surveyor, vii-x
Surveyor, measuring, viii-xi
Surveyor, quantity, vii, x, xi, xii, 27, 65
Surveys, land, 67; market, 66-68, 71
Systems, building, 14, 20, 31, 34, 53, 54, 125, 143, 145, 151; electrical, 18, 147; heating, 16, 128, 134, 139, 144, 152; internal communication, 73; mechanical, 18, 147; structural, 32

Target cost, 57-59
Target rates for internal rate of return, 106-108, 112-114, 123
Tax, advantages, 151; deduction, 123; implications, 95; legislation, 117; losses, 123; payable, 121, 123; rate, 117, 120, 122; rate, effective, 118; returns in Ontario and Quebec, 118
Taxable income, 118, 119, 121, 123
Taxation, 114, 117-124, 137, 150, 151
Taxes, 4, 72, 140; capital, 69; corporation, 117-118; federal, 117, 118, 120; improvement, 69; land transfer, 67; provincial, 117, 118, 120; real estate, 69, 72, 73; sales, 125
Taxpayers, 118, 120
Technology, changes in, 1, 141
Telephones, 70, 73
Temporary accommodation, 147; facilities, 15
Tenant finishes, 70; inducements, 70; lease takeovers, 70; partitions, 70
Tenants, 70, 72, 75
Tender, analysis, 6; coordination, 5; forms, 6; prices, 7
Tendering, ix, 5, 6
Tenders, x, xi, 2, 5, 6, 20-23, 27, 49, 59, 60, 62
Term of lease, 67; of mortgage, 76, 77
Terminal loss, 119

239

Terms of loan, 76; rental, 78
Testing of variables, 153
Theatres, 72
Time, 3, 50-51, 140-142; construction, 28, 131, 132, 143; and money, 80-93; period of an investment, 80-86, 88, 99, 103; of occupancy, 71, 130; when yield analysis is calculated, 116
Title insurance, 67
Total cost, 23, 65, 79, 98, 125, 145
Trade breakdown, 27, 28, 34, 59, 60; packages, 3, 22, 60
Trades, viii, x, 34, 44
Tradesmen, 9
Traffic, consultants, 65, 68; control, 15
Transfer costs, 96
Transportation facilities, 67
Trust deeds, 68
Turnover, in buildings, 141; of staff, 150
Type, of building, 13-15, 49, 54, 141, 143, 145; of construction, 29-31, 40, 44, 55, 145, 161; of occupancy, 14
Typical bay method, 32, 53

Ultimate costs, 125
Unambiguous drawings and specifications, 20
Unclaimed capital cost, 119
Underpinning, 15
Underwriters' fees, 68
Union contractors, 14
Unit cost, 18, 44; cost method, 29-31; in place method, 31, 32, 34, 58, 59; price, viii, xi, 1, 6, 15, 23, 25-28, 30, 31, 44, 48, 53, 56, 161
Upper floor, 18
UPPER FLOOR CONSTRUCTION, 35, 40, 44, 163, 177
Upright, viii
Usable space, 29, 30

Vacancies, 73
Vacancy allowance, 73; rate, 66
Validity of life-cycle costs, 151
Valuation, 4, 98
Value, ix; advertising, 126, 150; cost and, 3, 4, 65, 96, 97; created, 142; deferred, 84; of a development, 4, 66, 71, 96; of early occupancy, 143; future, 81, 82, 84, 91, 109, 130, 132, 133, 136; gain in, 126; of interior finishes, 54; of an investment, 95; land, 67, 98; market, 3; for money, 1, 48, 49, 60, 125, 153; of money, 80, 95; present, 84, 86, 96-100, 102, 104-106, 108, 109, 114, 117, 128, 130, 132-134, 136-141, 150; residual, 99; salvage, 139; theory of, 3
Variables, 152, 153
Variations, 50
VERTICAL MOVEMENT, 35, 54, 166, 183
Viability, 50, 65-67, 77, 99, 101
Vinyl-asbestos tile floor, 142
Volume of buildings, 30; of construction, 7

WALL FINISHES, 35, 54, 167, 186
Wall finishes, 54
WALLS ABOVE GROUND FLOOR, 35, 36, 53, 54, 57-58, 164, 180
WALLS BELOW GROUND FLOOR, 35, 36, 54, 163, 179
Walls, exterior, 16, 34, 54, 58, 72, 157-159
Warehouse, 18
Warranty, contractor's, 69
Washrooms, 72
Wastage (waste), 20, 25
Wasting asset, 86
Water, 73, 144
Wholesale trade, 13
Wind loadings, 18
WINDOWS, 35, 53, 164, 180
Working drawings, 21, 22, 49, 59; environment, 145
Workmanship, 25, 146
Workmen, viii, ix, 15, 18, 24, 25

Years' purchase, 88; with a limited term, 88
Yield, 75, 76, 94-96, 99, 101, 104, 106, 116, 117; analysis, 94-116; assumptions in calculating, 94-96, 116; calculations, 117; rate, 75, 94, 96-109, 113, 114, 116, 121, 122, 124